H. H. T. Jackson.

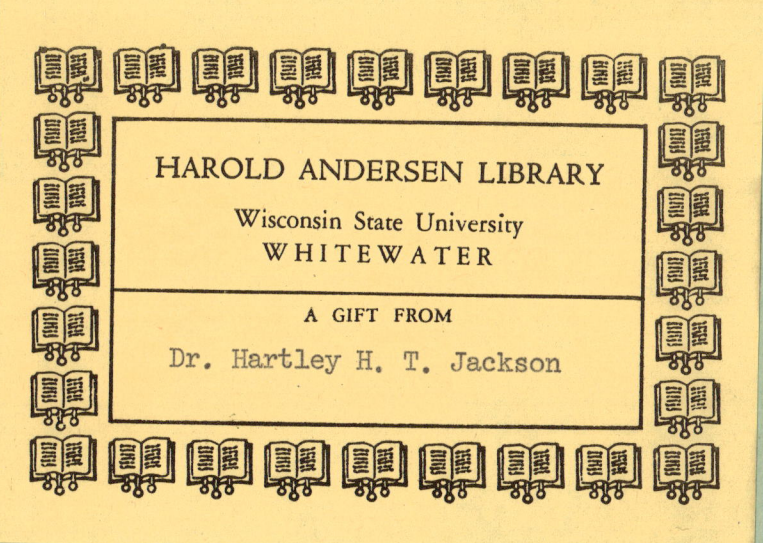

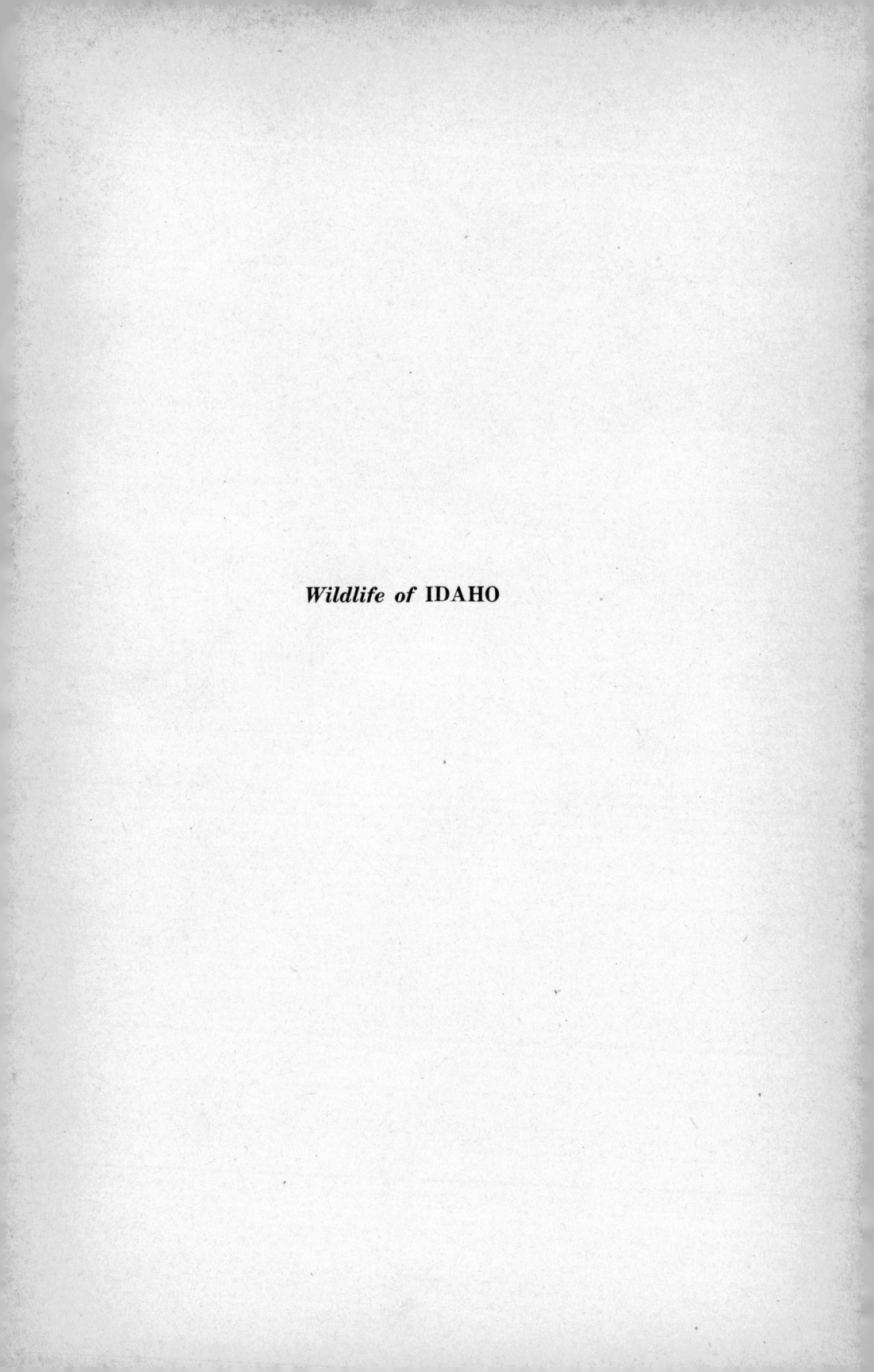

Wildlife of IDAHO

Mountain Goats

WILDLIFE
of
IDAHO

by

WILLIAM M. RUSH

FISH AND GAME COMMISSION
BOISE, IDAHO
1942

COPYRIGHT 1942 BY
IDAHO FISH AND GAME DEPARTMENT
BOISE, IDAHO

Printed and bound in the United States of America by
The CAXTON PRINTERS, Ltd.
Caldwell, Idaho
59009

To the Boys and Girls of Idaho:

Our Fish and Game Commission is giving you this book so that you may better know the mammals, birds and fish of Idaho.

Idaho is blessed with a great many forms of wildlife. It is rich in wild things as it is in minerals, forests, rangelands, cultivated fields, lakes and rivers.

The wildlife is yours. It belongs to the boys and girls of Idaho. The older folks are only taking care of it for you until you reach the age of manhood and womanhood when you will take over the reins yourselves. You must watch to see that these older people do not neglect your wildlife. You must guard against their killing too many of the game mammals and birds or catching too many fish. You must see to it that the wild creatures are well taken care of, that they have good homes with plenty to eat, whether it be in the forests, fields or waters. You should have more animals, rather that less so that there will be wildlife to pass on to your children and grandchildren.

Never kill any living thing just for the sake of killing, be it mammal, bird, snake, toad or fish. Harmful pests must be controlled, but many things considered harmful may be a great benefit to us. Be sure that a bird or mammal really is harmful before you kill it. Learn its habits and be sure.

By all means hunt. Learn to be a good hunter. Shoot and shoot straight. By all means fish. Learn to do it well. Above all, though, learn to be a good sportsman, and that means to be fair, honest and liberal with all the wild things. A good sportsman grows into a good citizen. He wants everyone else to love the outdoors as he loves it.

I sincerely hope you will enjoy studying this book and will profit greatly from it.

Chase A. Clark
Governor

STATE OF IDAHO

CHASE A. CLARK, Governor

DEPARTMENT OF FISH AND GAME COMMISSION

A. R. HOWELL, Chairman   G. W. GREBE, Secretary

W. A. FISCUS   G. E. BOOTH

A. L. TRADA

J. O. BECK, Director

DEPARTMENT OF EDUCATION BOARD

ASHER B. WILSON, President   J. H. ANDERSEN

W. F. McNAUGHTON   J. F. JENNY

MRS. A. A. STEEL

C. E. ROBERTS, State Superintendent

ACKNOWLEDGMENTS

Prof. H. M. Barr, Director of Research and Superintendent of the Irvington School in Portland, Oregon, furnished the author with advice and assistance of the utmost value in the selection and preparation of material suitable for grade school use. Much of the part on mammals was tried out in classes under his direction.

Miss Rita Hanson, Idaho State Elementary School Supervisor, read and checked all the material and passed on its suitability as to vocabulary and subject matter.

Texts consulted for technical information include:

Field Book of North American Animals. Anthony, H. E., 1928, Putnam

Birds of Canada. Taverner, P. A., 1934, National Museum of Canada

American Food and Game Fishes. Jordan and Evermann, 1934, Doubleday, Doran & Co.

Mr. Burton Perrine and Mr. Geo. Grebe of the Idaho Department of Fish and Game furnished technical and other information on Idaho fishes.

We are indebted for our illustration material to the following sources:

Dr. A. E. Weaver; U. S. Forest Service; Idaho Fish and Game Department; National Park Service; O. J. Murie; K. D. Swan; Lou Ovenden; Paul Sheffer; L. C. Todd, by Mr. Paul, Frontispiece; Osgood Smith; Ross Hall; Wisconsin Conservation Department; Ansgar Johnson; *Northwest Nature Trails*, Drawing of salmon by Quincy Scott.

A number of the bird pictures are of mounted specimens, the work of Oscar Jenkins. They are a part of the splendid bird display in the Capitol Building. Other photographs are by the author.

Grateful acknowledgment is also made to the many others who contributed in a large degree to the preparation of this volume.

W. M. Rush

Boise, Idaho, April 1, 1942

TABLE OF CONTENTS

	Page
NATURAL HISTORY—Chart	16
INTRODUCTION	17

PART ONE

MAMMALS

CLASSIFICATION OF MAMMALS—Chart	20
MAMMALS	21

Order Two

MEAT-EATERS (Carnivora)	25
DOG FAMILY (Canidae)	26
Wolves, Coyotes, Foxes	
Bow-Wow and His Relatives	27
CAT FAMILY (Felidae)	34
Mountain Lions, Lynxes, Bob-cats	
Reno, Custer and Darby	38
BEAR FAMILY (Ursidae)	48
Black Bears, Grizzlies	
Geolu	51
Geolu's Winter Sleep	55
RACCOON FAMILY (Procyonidae)	59
Raccoons, Ring-tails	
WEASEL FAMILY (Mustelidae)	60
Martens, Fishers, Minks, Wolverines, Otters, Skunks, Badgers	
Black-tip, the Killer	61
The Weasel	64

Order Three

INSECT-EATERS (Insectivora)	69
SHREW FAMILY (Soricidae)	69

Order Six

WINGED-HANDS (Chiroptera)	70
COMMON BAT FAMILY (Vespertilionidae)	70

CONTENTS

Order Seven

GNAWERS (Rodentia) — Page 75

SQUIRREL FAMILY (Sciuridae) — 78
- Woodchucks — 78
- Rock Squirrels — 78
- Say Ground Squirrels — 79
- Common Ground Squirrel — 80
- Western Chipmunks — 80
- Tree Squirrels — 81
- Flying Squirrels — 83

POCKET-GOPHER FAMILY (Geomyidae) — 84
- Western Pocket Gopher

RAT AND MOUSE FAMILIES — 86
- Non-native Rats and Mice (Muridae) — 86
- Pocket Rats, Pocket Mice (Heteromyidae) — 87
- Native Rats and Mice (Cricetidae) — 89
- Jumping Mice Family (Zapodidae) — 93

BEAVER FAMILY (Castoridae) — 95
- Beaver Ranch — 96

PORCUPINE FAMILY (Erethizontidae) — 109

Order Eight

RABBIT-LIKE (Lagomorpha) — 111

PIKA FAMILY (Ochotonidae) — 111
- Little Chief Hare — 112

HARE AND RABBIT FAMILY (Leporidae) — 115

Order Nine

HOOFED (Ungulata) — 120

DEER FAMILY (Cervidae) — 121
- Moose, Caribou, Elk, Deer
- A Deer Hunt — 132

PRONG-HORN FAMILY (Antilocapridae) — 135
- Antelopes
- How Antelope Horns Grow — 137

HOLLOW-HORN FAMILY (Bovidae) — 138
- Cattle, Sheep, Goats
- Buffalo — 138
- The Year of the White Buffalo — 140
- Mountain Sheep — 147
- Mountain Goats — 149
- Early Days in Idaho — 150
- Rules of the Game — 161
- Counts, Estimates and Guesses — 167

CONTENTS

PART TWO
BIRDS

		Page
BIRDS		173
CLASSIFICATION OF IDAHO BIRDS—Chart		175

ORDER

One	Loons	(Gaviiformes)	176
Two	Grebes	(Colymbiformes)	176
Three	Pelicans	(Pelicaniformes)	178
Four	Herons, Storks, Ibises	(Ciconiiformes)	178
Five	Ducks, Geese, Swans	(Anseriformes)	180
	Wild Geese Are Calling		185
Six	Vultures, Hawks, Eagles	(Falconiformes)	194
	Pandion, the Osprey		196
Seven	Grouse, Pheasants, Quail	(Galliformes)	202
	Opening Day		213
Eight	Cranes, Mud Hens, Coots	(Gruiformes)	220
Nine	Shore Birds	(Charadriiformes)	221
Ten	Pigeons, Doves	(Columbiformes)	223
Eleven	Owls	(Strigiformes)	224
Twelve	Goat-Suckers	(Caprimulgiformes)	225
Thirteen	Swifts, Hummingbirds	(Micropodiformes)	226
Fourteen	Kingfishers	(Coraciiformes)	226
Fifteen	Woodpeckers	(Piciformes)	227
Sixteen	Perching Birds	(Passeriformes)	227

IDAHO STATE BIRD - 232

PART THREE
FISHES

	Page
FISHES	235
CLASSIFICATION OF IDAHO FISHES—Chart	241
CLASSIFICATION	242
ORDER OF SPINY-FINNED FISHES (Acanthopterygii)	243
Bass and Sunfish Family	244
Just Fishin'	248
ORDER OF BACK FINS AND FAT FINS (Isospondyli)	254
Trout Family	255
Blackspot and Doll	261
Salmon Family	267
Salmon Fishing	269
A Thousand Miles Down to the Sea	273
Whitefish	279
Grayling	280

CONTENTS

	Page
ORDER OF SUCKERS AND MINNOWS (Plectospondyli)	282
Suckers	283
Minnows	284
ORDER OF CATFISH (Nematognathi)	288
ORDER OF SCULPINS (Scleroparei)	291
ORDER OF STURGEONS (Glaniostomi)	292
Rearing Young Fish	294

LIST OF ILLUSTRATIONS

	Page
Mountain Goats, *Frontispiece*	Facing 3
Coyote	27
Bob-cats	36
Grizzly Bear	48
Black Bear and Cubs	50
Geolu	51
Ring-tail	59
Black-tip, the Killer	61
Mink	65
Spotted Skunk	67
Woodchuck	79
Ground Squirrel	79
Western Chipmunk	81
Pine Squirrel	81
White-footed Mouse	89
Neotoma	91
Muskrat	93
Beaver, chewing Green Aspen	95
Dam the Beavers Built	96
Beaver in Live Trap	103
Snowshoe Rabbit	115
Cottontail Rabbit	118
Moose	124
Elk	126
Baby Elk	128
White-tailed Deer	129
Black-tailed Deer	130
Antelope—*Color*	Facing 135
Buffalo	140
Mountain Sheep	147
Grebes	177
Blue Heron	179
Mallards—*Color*	Facing 181
Mergansers	182

ILLUSTRATIONS

Page

Cinnamon Teal Ducks	182
Melander and Branta	185
Goshawk	194
Sparrow Hawk	194
Ospreys	196
Blue Grouse	202
Ruffed Grouse	205
Sharp-tail Grouse	205
Sage Grouse	206
Ptarmigan	207
Ring-necked Pheasant—*Color*	Facing 207
Hungarian Partridges	209
Chukar Partridges	209
Bob White Quail	210
Valley Quail	211
Mountain Quail	211
Opening Day	214
Coot and Mud Hen	220
Avocets	222
Mourning Doves	224
Short-eared Owl	225
Belted Kingfisher	226
Flicker	227
Pewee	228
Flycatcher	229
Song Birds	230
Song Birds	231
Diagram of Fishes	236
Smallmouth Bass—*Color*	Facing 244
Largemouth Bass—*Color*	Facing 244
Bluegill Sunfish—*Color*	Facing 246
Crappie, or Calico Bass—*Color*	Facing 246
Yellow Perch—*Color*	Facing 247
Just Fishin'	249
Rainbow Trout—*Color*	Facing 255
Brown or Lockleven Trout	256
Mackinaw Trout	257
Eastern Brook Trout—*Color*	Facing 259
Dolly Varden Trout—*Color*	Facing 260

ILLUSTRATIONS

Page

Cutthroat Trout—*Color*	Facing 261
Blackspot and Ruby in Nest	263
Little Redfish, Sockeye or Blueback Salmon—*Color*	Facing 267
Steelhead	268
Chinook Salmon	271
Whitefish	279
Grayling	280
Sucker	283
Squawfish	284
Carp	285
Shiner	286
Chub	286
Bullhead	288
Channel Catfish	289
Sturgeon	293

NATURAL HISTORY

KINGDOMS   BRANCHES   CLASSES

I. ANIMAL—Zoology

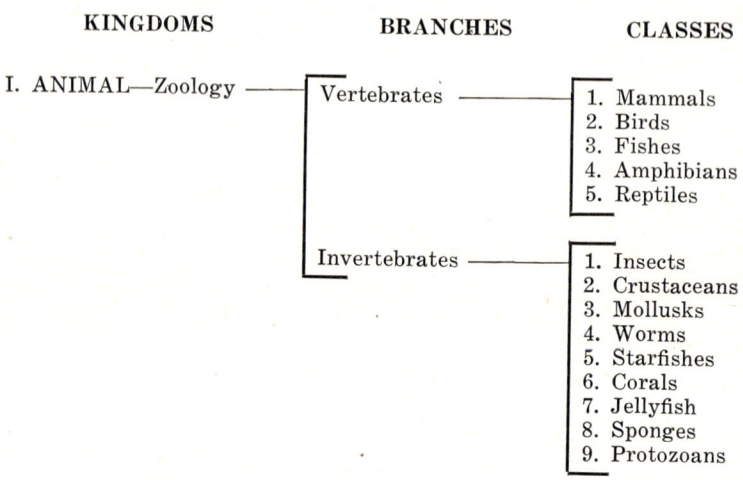

    Vertebrates
1. Mammals
2. Birds
3. Fishes
4. Amphibians
5. Reptiles

    Invertebrates
1. Insects
2. Crustaceans
3. Mollusks
4. Worms
5. Starfishes
6. Corals
7. Jellyfish
8. Sponges
9. Protozoans

II. PLANT—Botany

    Examples

- Trees
- Shrubs
- Grasses
- Weeds
- Cacti
- Mosses
- Toadstools

III. MINERAL—Geology

    Examples

- Gold
- Granite
- Soil
- Water
- Iron
- Sand
- Coal

INTRODUCTION

NATURAL HISTORY is the study of natural objects around us. It is divided into three great kingdoms: **animal, plant** and **mineral.**

Zoology deals with all animal life, from tiny bugs which can be seen only through a microscope, to angleworms, birds, fishes, snakes, dogs, cattle, elephants and men.

Botany deals with all plant life, from toadstools, moss, grass, weeds, to rose bushes and giant fir trees.

Geology deals with all things that do not live, grow and reproduce their kind. It is a study of the earth's rocks, soils and minerals. Thus, water belongs to the mineral kingdom, as well as do iron and clay.

This volume treats of the wild animals, birds and fishes of the State of Idaho; therefore, it is a study that comes under the heading of the **Animal Kingdom,** or **Zoology.** The first division of zoology is into two great groups: VERTEBRATES, animals with a backbone; and INVERTEBRATES, animals without a backbone. It is easily seen that such animals as dogs, birds, fishes and snakes are vertebrates, while oysters, angleworms, butterflies and lobsters are invertebrates.

Vertebrates are divided into five *Classes*, as follows:

Mammals are warm-blooded animals which suckle their young. The word mammal comes from the Latin "mamma," in English "teat" or "tit." Examples of mammals are: sheep, rabbits, mice and deer.

Birds are warm-blooded animals covered with feathers, that lay eggs from which their young are hatched. Robins, turkeys and ducks, of course, come under this heading.

Fishes, cold-blooded vertebrates, usually covered with scales, live their whole lives in water. They breathe by means of gills and their young are born from eggs. Trout, cat-fish and bass are examples. (A few species of fishes bear their young alive, instead of laying eggs.)

Amphibians spend part of their lives in water, but later

become air breathers and live, at least part of their lives, on land. They, too, are cold-blooded, like the reptiles. In this class are frogs, toads and salamanders.

Reptiles are cold-blooded. Snakes, lizards, turtles and alligators belong to this class. Some reptiles give birth to live young, while others, like birds, lay eggs.

PART ONE

MAMMALS

CLASSIFICATION OF MAMMALS

CLASSES	ORDERS	FAMILIES	EXAMPLES
1. Mammals	1. First *Primata*	Human Monkey	Man Ape
2. Birds 3. Fishes 4. Amphibians 5. Reptiles	2. Meat-eaters *Carnivora*	Dog Cat Bear Raccoon Weasel	Coyote Mountain lion Grizzly Coon Weasel
	3. Insect-eaters *Insectivora*	Mole Shrew	Mole Shrew
	4. Fin-footed *Pinnipedia*	Eared seal Earless seal Walrus	Sea lion Harbor seal Walrus
	5. Whales *Cetacea*	Baleen Sperm Dolphin and porpoise Fresh water	Blue whale Sperm whale Killer whale
	6. Wing-handed *Chiroptera*	Bats—several families	Brown bat
	7. Gnawers *Rodentia*	Squirrel Pocket gopher Rat and mouse Beaver and mountain beaver Porcupine	Chickaree Gopher Wood rat Beaver Porcupine
	8. Rabbits *Lagomorpha*	Pika Hare Rabbit	Pika Jack rabbit Cottontail
	9. Hoofed *Ungulata*	Deer Antelope Cattle, sheep, goats	Moose Pronghorn Buffalo
	10. Pouched *Marsupialia*	Kangaroo Opossum	Kangaroo Opossum

MAMMALS

Part One of this book is about the first class, *mammals*. There are so many of them and they differ so widely in form and habits that they are divided and classified in *orders*.

The first order is made up of MONKEYS, APES and BABOONS. These are mammals which have hands and hand-like feet well adapted in most species to climbing trees and walking upright. People belong to this order, too, although many of them do not like to be reminded of it.

The second order is the MEAT-EATERS; dogs, cats and other animals whose diet is chiefly made up of other animals.

Third order is the INSECT-EATERS, the moles and shrews, which are rather small mammals which live most of their lives underground and eat insects.

Fourth order is the FIN-FOOTED animals which live most of the time in the ocean. Instead of having hands and feet to walk on, they have fins which help them to swim. Sea-lions and seals are examples of this order. They are covered with fur for warmth.

Fifth order is the WHALES, huge fish-like mammals without hair. They spend their entire lives in water. A thick coat of fat, or blubber, under their slick skins, gives them warmth. Whales are much larger than any of the other mammals.

Sixth order is the WING-HANDED. These mammals fly like birds.

Seventh order is the GNAWERS. Rodents, we call them, and they include squirrels, woodchucks, chipmunks, mice, rats, beaver and porcupines. These have long sharp front teeth for gnawing bark.

Eighth order is the RABBITS. Snowshoes, cottontails, conies and jack rabbits belong to this order.

Ninth order is the HOOFED animals, such as cows, sheep, pigs, deer and antelope.

Tenth order is the opossums and kangaroos, mammals which carry their young in a pouch. POUCHED animals, we call them.

It might seem that division of the animal world had gone

far enough. Not quite, however—there are a few more to learn. Orders are divided into **FAMILIES**. For instance, the meat-eaters are divided into five families; bear family, raccoon family, weasel family, dog family and cat family.

A **species** is the smallest sub-division in dealing with the wild mammals, except, of course, any one animal, which is an *individual*. We will use the word species quite often as we study wild animals.

The animals most familiar to us are individuals. Dreamy-eyed old Tom, sleeping on a cushion by the fire, is, first of all, "our old cat." He is an individual, different in a good many ways from any other cat on earth. His eyes are a little more yellow, his fur softer, his voice louder, perhaps. He has his own way of pouncing on a mouse, his teeth are sharper than those of other cats. Yet, he is enough like the ordinary house cat to be a member of the species, "house-cat."

There are other species of cats. There are the lynx, bobcat, mountain lion, tiger and leopard, which make up the cat family. They are described as: "Furred mammals of small to large size, round head, walk on their toes, five toes on each front foot, four toes on each hind foot, sharp claws that can be pulled into a sheath, teeth of the shearing type for tearing and cutting meat, a very rough tongue, and, finally, the ability to climb trees."

The cat family are meat-eaters, which puts them in the **ORDER** of meat-eaters, with dogs, martens, bears and coons. They are warm-blooded animals which bear their young alive and nourish them at the breast. Together with the other *orders*, they make up the *class* of mammals, which brings us back to where we started on Page 1.

Just a few words about families. Our own family at home is father, mother, brother and sister. It is a very nice family of four people. What about grandfather and grandmother who came to visit last Christmas? They are part of our family, too. So are Uncle Amos and Aunt Hester. Mother has aunts and uncles, besides three sisters and a brother, whose children are our cousins. They are part of our family, of course. How about second cousins, third cousins and father's nephew's wife's sister's husband? Yes, a little distant, but still our family.

Come to think of it, all people on earth are in our family, the

HUMAN FAMILY. "People," we call them, or "folks," if we are feeling especially friendly toward them. *Mankind* is another word we use. Sometimes we simply say *man,* and mean every person on earth, Chinese, Malays, Negroes, Whites and Indians. They are the yellow, brown, black, white and red races of beings. All of them together make up one *family* in the animal kingdom.

So, when we speak of the dog family we do not mean just papa dog, mamma dog and the puppies. We mean all of the dogs on earth. Wolves and coyotes, too. The foxes are sort of second cousins to the dogs, so we put them in the dog family. If we want to speak of papa dog, mamma dog and the puppies as a family, we say, "Old Rover's family," or "Scotch's family." The dog family means all dogs. Rover's family means just his own little group.

Likewise with the cat family, the deer family and all the rest. It is very important in the study of wild animals to know to what family any individual wild animal belongs.

Why have Latin and Greek names for plants and animals?

Latin is used the world over by scientists of all nationalities. It is a sort of universal language. It has the advantage of being a dead language and not subject to growth and change, as English is. For example, *catus* in Latin will always mean "cat" just as it did hundreds of years ago. In English, "cat" has many meanings, a few of which are: caterpillar tractor, cat-o'-nine tails, a spiteful female, and catfish. And, as time goes on, there will be more meanings. The Latin name means just one thing and is fixed for all time.

Latin to the scientific world is like the musical score to the musical world, understood everywhere.

Many Greek names have been in use for many hundreds of years. It would be foolish to change them now.

Thousands of forms of vertebrate life are not mentioned in this book. Orders of Mammals I, IV, V and X are left out. You will find nothing about reptiles and amphibians. Nor is there anything about the invertebrates. Besides hundreds of thousands of species of insects; ants, flies, mosquitoes, potato bugs, honey bees and butterflies; there are thousands of species of worms, shellfish, starfish, jellyfish, corals, sponges and tiny animals too small to see without a microscope. All these make up the animal kingdom of several hundred thousand species.

Some of these will come up for future study in high school and college.

Men and women, girls and boys, live in this same world and share its air, soil, water, trees, grass and other good things with these animals. They are a part of our life and we are part of theirs. Unless we know these wild animals rather well, we will miss much enjoyment.

We very diligently provide food and shelter for our animal pets and farm animals. None of us would think of leaving our dog to find food for himself in the woods on a winter day, nor do we make the milk cows rustle their own forage. The wild things have to provide for themselves, though, and they do it very well, each in his own way. People can learn much from them.

ORDER OF MEAT-EATERS

(CARNIVORA)

DOG FAMILY	(Canidae)
CAT FAMILY	(Felidae)
BEAR FAMILY	(Ursidae)
RACCOON FAMILY	(Procyonidae)
WEASEL FAMILY	(Mustelidae)

Carnivora is Latin for "flesh-devouring" family.

Meat-eaters have powerful teeth and jaws. Some of them have long, sharp claws. All are ferocious and merciless when hungry and hunting food. In this order are the most intelligent of all the wild creatures in the animal kingdom. Some of the animals of this order cause man a great deal of trouble by killing horses, cattle, sheep and poultry. Some are valuable for their furs. Man has obtained his best four-footed friend and helper from the dog family. The cat family has given us some fine pets.

DOG FAMILY

(CANIDAE)

Canidae is Latin for "dog-family." There are a number of important members—wolves, coyotes and foxes, besides a great number of domestic species.

In one way and another the dog family is the most important one to people, of all the wild animals. Some are very destructive. Some are valuable in keeping a check on rabbits, ground squirrels and mice. Some are good friends of man and work willingly for him. Some have valuable skins. To a boy or girl there is not a puppy of any kind on the face of the earth that he or she would not like to have for a pet, be it wolf, coyote, dingo or fox. Of all the animals, dogs have always been the best friends of man.

BOW-WOW AND HIS RELATIVES

Do you know what a turn-spit is?

If you had lived five thousand years ago, you might have owned a long-bodied, crooked-legged dog that was called a turn-spit. He came quite naturally by that name because he was a dog that turned the spit.

COYOTE

In those days, cooking was done over an open fire. There were no stoves. Meat was cooked in large chunks rather than sliced and fried or chopped up into stew. It was cooked on a spit. A spit is nothing more than a long wooden rod which was pushed through the piece of meat to be cooked. Then it was suspended over the fire on forked upright sticks driven into the ground on each side of the fire. In order that the meat be cooked evenly on all sides, the spit was slowly turned, over and over, for maybe two hours.

How in the world could a dog turn the spit?

The people in those days were pretty good mechanics. They did not have gasoline or electricity, but they had machines for developing power, which were called tread-mills. Some of the mills were large ones in which a dozen or more men worked. These developed a lot of power. Others were small in which only one dog worked. These were used to turn the spit for cooking, so that the cook could devote all of his time to tending the fire, doing other cooking, or just loafing in the shade. Turn-spits were slow, heavy-bodied short-legged dogs.

Five thousand years ago was a long time back. There were in those days, besides turn-spits, such other breeds of dogs as greyhounds, house dogs, toy dogs, and, of course, curs.

If you should go to a dog-show in any large city, you would find that there are now about two hundred breeds of dogs. They are classified into six groups: sporting dogs, hounds, working dogs, terriers, toy dogs and non-sporting dogs. All of these are descendants of wild dogs, modified by being tamed and living with people.

The head of the wild dog family is the **wolf.** He is a big grey fellow, four feet long, and has a beautiful bushy tail about a foot and a half in length. He may weigh as much as a hundred and twenty-five pounds. Quite often there are wolves of other colors than grey. Some packs have a pure white or a coal black wolf running with them.

The wolf eats any animal that it can catch and kill, from mice to moose, including birds and fish. It hunts both by day time and by night. Sometimes wolves work in pairs or small packs to run down deer, cattle or colts. Wolves are savage and very smart. A few of them can do a great deal of damage to sheep and other domestic animals. Men use rifles, traps and poison to kill every wolf they can find. In spite of this, there are still wolves in the mountains of northern Idaho and in other parts of the United States, Canada and Alaska.

This big wild dog is one of the ancestors of some of the breeds of our tame dogs. Perhaps you have seen sled dogs, such as malemutes, huskies or eskimos and said, "Why, that looks just like a wolf!" They do have some wolf blood in them. Some breeds used in the Far North are half wolf.

Wolf pups are easy to tame, but when they grow up they become very cross and dangerous.

In India there is a smaller wolf about the size of our coyote, which has a very bad reputation for killing children.

A species of wolf that has almost disappeared from the earth is the **maned wolf** of South America. He used to be found in Brazil, Paraguay and Uruguay. He hunts at night, all by himself, and is a very fast runner. He is about the size of a coyote, but has very long legs and a stiff, coarse mane on his neck. He eats small mammals, insects and fruit. He is easily tamed, even after he is full grown and he has a very good disposition. The maned wolf, or one much like it, may be the ancestor of our greyhound.

The best known of all the wolves is, of course, our American **coyote.** Its Latin name is *canis latrans* which means "barking dog." Probably every Idahoan has seen one of them at some time in his life. The coyote is a beautiful grey in color, about three feet long, and has a large bushy tail a foot and a half in length. The average Rocky Mountain coyote weighs about thirty pounds and has a fine coat of dark, silky fur that is almost as fine as that of a fox. In the southern states, coyotes are smaller, not as pretty and have shorter fur.

Coyotes, like all wolves, prey upon other animals. Rabbits are their most important food; then come small rodents, such as mice and ground squirrels. They also eat carrion, which means animals they find dead. Of course, they kill some deer, antelope, birds, chickens, sheep, calves, pigs and goats. Often they are blamed for killing more than they actually do.

The coyote is a very fast runner, a good fighter and has a very fine imagination when it comes to outwitting a trapper or hunter. The mother coyote takes excellent care of her pups. She is clever at picking out good den sites and in getting food for her young ones. In spite of all men do to get rid of coyotes, we have more of them every year. They have spread over nearly all of the United States and gone far into Canada.

It is easy to tame a coyote. Indian tribes cross-bred their camp dogs to coyotes in order to get a finer coat of fur. Some of our domestic dogs may have coyote ancestors to thank for much of their intelligence, physical strength and good dispositions. Have you ever noticed how much a Belgian shepherd dog looks like a coyote? It is hard to tell them apart if they are playing a little distance from you.

If we went over to South Africa, we would find a wild dog about the size of a coyote. It is called the **black-backed jackal** and is so destructive to game, poultry and domestic animals that there is a big bounty paid for every tail brought in. It is a much worse nuisance than the coyote.

In Australia there is a wild dog called the **dingo.** It is a little bigger than a coyote and much more destructive to sheep and poultry than any coyote that ever lived. Some people think that the dingo is a tame dog that has gone wild. They think its ancestors were brought into Australia and turned loose to forage for themselves.

In India there is a red dog called the **dhole.** Kipling wrote a great deal about them in his *Jungle Books*. They are almost as big as Arctic wolves and hunt in packs of twelve to thirty. They eat mostly antelope and deer, hunting behind a leader whom they obey most willingly. They do not bark or howl. They hunt by *scent* and herd the deer or antelope close to their dens before killing it, so they have food near at hand for their puppies. Often many dhole dens are close together. The pups are easily tamed, but get cross and dangerous when they are grown.

What tame dogs can you think of which might have descended from the dhole? Remember his red color, how he hunts by scent, how intelligent he is and what his character is.

In South America we find another wild dog, called a **raccoon dog,** because it looks something like a coon. It "plays possum" when caught and is as harmless as a poodle or spaniel. It lives on mice, lizards, and—of all things—sugar cane!

Then there is the **hunting dog** of Africa, to be found south and east of the great Sahara desert. This wild dog weighs 75 pounds and is 26 inches tall, with long legs, short nose and large, erect ears. His coat is spotted and blotched yellow, brown, black and white.

These dogs hunt in packs, trail by scent, and never bark or howl when hunting. When they sight their prey, they close in and, since they are said to be faster than the African antelope, they have little trouble running down any animal they choose. They are very ferocious. When other food is scarce they kill hyenas, and they will even tackle a lion. Any pack of animals that will try to kill a lion must be almost the bravest animals in the world. We can easily see some of the good points of the African hunting dog in our domestic dogs.

Dogs love to run in packs under a leader. They pick one of their number as captain and obey him without question. Dogs are great for team work. Let us see how two coyotes team up to kill a young antelope kid.

The mother antelope is cropping grass and nibbling at the tender new growth of silver sage way out in the Owyhee country. Her three-day-old kid is curled up in the bunch grass and sage fifty yards away.

The coyotes are on a rise in the ground two hundred yards away. They have watched her nurse her kid and seen it curl up in the grass. They watch the old antelope as she grazes. Every few seconds she raises her head and scans the country carefully for possible enemies. The "barking dogs of the plains" have a very healthy fear of her sharp hoofs, so they wait and plan their attack carefully.

One coyote circles to the right, the other to the left. They are close when the antelope sees them. She starts for one, stops, turns around and starts for the other. She is undecided. She stops again. The coyotes come nearer. Mother antelope dashes madly at one of them, her front feet ready to cut it down. Coyote number two rushes in toward the spot where the kid is lying. The old antelope looks back over her shoulder and sees her young one in danger. She whirls and dashes back. Number two leads her as far away as he can. Number one slips in, a grey streak of hungry dog, bent on making a kill. It reaches the antelope kid and slashes its throat before the old antelope can get back. She chases it furiously, but it is too late. Both coyotes move away and wait. Mother antelope finds her kid dead and no longer tries to protect it. She goes away to graze and the barking dogs move in to their feast.

This is a sample of dog team work.

Now, let us see how coyotes kill deer in winter: The pack spreads out along a steep, icy hill with an open river at the bottom. There are a dozen or more coyotes this time, instead of just two. Two or three start the deer running down hill. A few are scattered along the way to jump out and keep the deer running its fastest. The deer sees the water that means safety and puts on an extra burst of speed. There is a slippery place where it slips and falls. One of the coyotes darts in to slash at its legs or throat. The deer scrambles to its feet and hurries on. There is another icy spot, another bad fall, another coyote there with wicked teeth. Finally the deer nears the water, but it is almost exhausted and waiting coyotes fall upon it to make the kill.

No other wild animals work schemes as clever as that one.

When men turn their tame dogs out, they revert to the wild in a very short time. Wild dogs are a menace to all kinds of smaller animals and birds. Some become terrible deer killers. In Idaho, in fact in all of the West, sheep-killing dogs do much damage. They are clever, cruel, silent killers, much worse than coyotes. Coyotes usually get blamed for all sheep killed by wild dogs.

Dogs do not bark when they go back to the wild. They howl, more or less as their ancestors did. Barking is an acquired habit that wild dogs do not have. Wolves, coyotes and other wild dogs howl, cry or yap when they want to make a noise, but they never bark as your neighbor's collie or feist so often does.

Foxes are not ancestors of any domestic dogs, although some of our dogs are a little fox-like in appearance and habits. Foxes are sly and secretive. They are cunning and resourceful when pursued by dogs or hunters. They have extremely sensitive senses of hearing, sight and smell. Foxes are the most beautiful members of the whole dog family. There are several species in North America.

Around the North Pole is the **Arctic fox.** It ranges the ice floes, tundras and open lands, south to the great forests. As one might expect of a fox that lives in a land of snow and ice, it is pure white in winter, a trifle brownish or slate colored in summer. This fox shows great intelligence in getting food. Often it follows a polar bear. The bear kills seals. The fox feasts on what the bear leaves. Where lemmings (a sort of Arctic mouse) are found, the fox kills hundreds of them and stores them away for food. It also stores away any other food it can find, such as fish. Arctic foxes readily kill and eat other foxes if they find them caught in traps. Trappers are enraged by such habits and call them cannibals. White fox skins are worth a hundred dollars or more apiece.

Sometimes the white fox has a bluish color. It is smoky grey, even a little reddish brown on the head and feet. A fine blue Arctic fox may bring the trapper two hundred dollars.

Farther south in Canada and over most of United States the **red fox** roams. This is one of the species found in Idaho. He is a beautiful golden red in color. In dry desert regions his color is pale. In Alaska he is brightest in color. The largest and most valuable skins come from the Far North.

Another color phase is the **cross fox.** This is a red fox marked with black legs, black tail and a dark-colored cross on the shoulders, also found in Idaho.

Then, there is the **grey fox,** that lives south of Canada. He is a big grey and black fellow with longer legs than the red fox has. His fur is inferior to that of other foxes, except the kit fox, a very small grey fox. Both of these species are found in Idaho.

Foxes sometimes do quite a lot of damage to flocks of poultry and small game animals. In many localities they are hunted down and killed because of the damage they do. Trappers have taken most of the foxes with valuable skins. It is said by experienced trappers that coyotes are natural enemies of foxes and drive them out wherever the coyotes go.

There are many fox farms in the United States, Canada and Alaska, where foxes are reared for their fur. There are some fox farms in Idaho. Many of the furs produced on these farms are made into ladies' coats and expensive neck pieces which find a ready market in the best stores.

CAT FAMILY

(FELIDAE)

Felidae is the Latin term denoting the cat family.

More than four thousand years ago cats were sacred animals in Egypt and to kill one, even by accident, was punishable by death. In an Egyptian household the cat was the most valuable possession. From Egypt cats were taken across the Mediterranean Sea to Europe and the British Isles. A thousand years ago cats were so valuable in England that a "fine of corn" was levied against anyone who killed one of them. The dead cat was suspended by the tail with its nose just touching the floor and the offender was compelled to heap grain high enough to cover the tip of the cat's tail. Cats were insurance against mice and rats.

House cats today are valued as pets in towns and cities. Many country dwellers keep them for mousers. Cats do not care to eat mice if they can find other food. They catch mice in a spirit of sport. If well fed, cats will catch more rats and mice. Poor, half-starved cats are not good mousers.

Egyptian cats became mixed with native wild cats in England, and also with tame cats from Russia, China, Japan and the Malay States. One thing about Egyptian cats has persisted through the centuries. Look at the soles of your old cat's paws. Are they black? If so, then he is of Egyptian descent. If brown, he is probably of Siamese descent.

The **Manx cat,** from the Isle of Man, is probably from Japan or China, by way of Russia. Why do we think so? The Manx cat has no tail, and the only bob-tailed wild cats in the Old World are found in the Far East. All other members of the cat family have long tails except our own lynx and bob-cat. Native American cats have had no part in the development of house cat breeds. Most of the cats we see roaming the city streets and country fields are mongrels, mixtures of many breeds.

Let us compare cats with dogs as pets. Cats are cleaner and fussier about their food than dogs. They keep their bodies cleaner, too. A cat's tongue is a very efficient hair brush and every tabby spends much of its time smoothing its coat.

Cats are not as noisy as dogs, nor are they as destructive to gardens and flower beds. They are not as ready to fight as the average dog, either. Cats are more independent and self-reliant than dogs.

Cats lack the dog's sympathetic understanding, however. Imagine, if you can, a cat trying to help someone in distress. Cats are not faithful. They love the *place* where they live, rather than the people who take care of them. Dogs love their masters and go willingly to the ends of the earth with them. Cats do not like to obey commands, whereas dogs enjoy doing it. Cats cannot be trained like dogs to do useful work such as herding sheep, driving cattle, carrying first-aid kits in battle, pulling sleds, running errands and hundreds of other things. The voice of the dog is very much the better of the two. As hunters, only the cheetah (and it really is almost as much a dog as it is a cat) has been trained to hunt with man.

Cats of all kinds, from the lordly lion to the smallest house cat, hunt by sight rather than by smell. Cats do not track their prey. They lie in wait and pounce upon it with powerful teeth and claws. Even old Tom is a good hunter when it comes to killing mice and birds for sport. His love for killing indicates that, in spite of his fondness for being petted, his warm hearth and meal of bread and milk, he is yet only half tame. Much damage is done to small mammals and birds by wandering house cats.

The **wild puma**, or cougar, of South America, is said to be friendly to man. Perhaps there is a strain of friendliness in all wild cats, but if there is, it is well hidden in most species. One can hardly see how a lion, tiger or leopard could be called friendly to mankind.

There are long tails and short ones in the cat family. Long-tailed members are: lion, tiger, leopard, mountain lion, jaguar, serval, ocelot, European wild-cats and most varieties of house cats. (The Manx cat we mentioned as an example of short-tailed house cat.)

Of the two species of short-tailed native cats, the **Canada lynx** is the largest and most beautiful. This is the lynx of Hudson's Bay fame, very much sought by trappers. Its range is

BOB-CATS—mounted

south from the Arctic to the Great Lakes, along the Rocky Mountains to Colorado and in the Cascades to Oregon. Canada lynx are rather scarce in Idaho. The lynx has a body length of nearly three feet, tail only *four inches* in length, and an enormous, hairy hind foot nine inches long. On each ear is a large tuft of black hair. The tail is tipped with black. Rabbits are its principal food, especially in the Far North. When rabbits are plentiful, trappers rejoice, for that means plenty of valuable lynx to trap. Lynx are able to travel well on top of snow as their very large furred feet act as snowshoes.

A close relative of the Canada lynx is the ordinary **American bob-cat.** This cat is somewhat like the lynx, but lives farther south from coast to coast, in mountains, foothills, settled communities, prairies and deserts. It is an able hunter. Where numerous it is a threat to all animals the size of deer and smaller. Chicken houses and sheep folds are never safe from this killer. Its fur is not valuable like the lynx's, but because of its destructive habits, the bob-cat is classed as obnoxious and a price is put on its head.

Bob-cats have small ear tufts. Their tails are longer, their hind feet shorter than those of the lynx. One certain way to tell a bob-cat from a lynx is by their tails. The lynx tail has a black tip *all around*, while a bob-cat's tail is tipped with black on top, *pale underneath*.

All wild cats are very secretive. They are active mostly at night. Dogs are their natural enemies. The largest and most ferocious mountain lion lives in mortal terror of even an ordinary fox terrier. Lynx and bob-cats are easily hunted down with dogs. Wild cats of some species or another are found all over the world except in Madagascar and Australasia.

Mountain lions have the imposing Latin name of *Felis concolor hippolestes,* in English, "colt-killing-grey-cat." Other names are: cougar, puma, panther and painter. In Idaho this cat grows to a body length of nearly six feet, plus a tail three feet long. It weighs as much as two hundred and twenty pounds. Its home is in the forests and mountains.

RENO, CUSTER AND DARBY

"Ting-a-ling! Ting-a-ling! Ting-ling-ling-ling-ling!" shrilled the telephone bell in George Lowe's comfortable home one bitter cold February morning.

"Two shorts and a long," said Mrs. Lowe. "It's not a call for you, thank goodness. This would be a terrible day to start out to the mountains."

"I'm satisfied to stay right here by the fireplace," said Mr. Lowe. "The way I feel right now I wouldn't care if I didn't have to go out again for a month."

"Well, I should think so! You have a hundred and sixty lions to your credit now. Isn't that enough for any man?"

"OO-oo-oooooow!" "Wow-ou-ow!" came from the shed behind the house.

"Listen to Darby and Reno answering you," Mr. Lowe laughed. "That sounds ridiculous to them."

"Boo-oo-ou!"

"And there's Custer, putting in his say-so."

"Those dogs know what we're talking about most of the time," said Mrs. Lowe.

The dogs' voices sounded louder. "Ba-wow! Ou-u-u! Woou-u-ow!" All of them spoke at once.

"They know more than that," said George. "Do you know what they're crying now?"

"They surely don't want to go lion hunting again so soon. Why, you just got in from a three weeks' trip to the Salmon river country."

"Good dogs always want to hunt. They can smell a hunt coming. That's what they're crying about right now."

"You can't go out in this bitter cold. I won't have it, lions or no lions! Why, you're not even good and warm yet, and you haven't had nearly enough to eat."

"Just the same, lions are killing deer somewhere. Reno and Custer know it and are anxious to get on their trail. Listen to that!"

There was a note of excitement in the hounds' deep-throated cries. They wanted to get out in the storm and track down their age-old enemy, the big mountain cat, that lived by killing deer.

Was it some old, almost forgotten urge that stirred the dogs to action this cold morning? Was it something that came to them on the icy wind that roared out of the Northland? Could they hear Mr. and Mrs. Lowe talking and understand what they said? Did they dislike cats so much that they were always willing to chase the big mountain lions, no matter how cold the weather, or how tired they got? The dogs suffered from cold. They got terribly hungry, too, at times. Right now they were so gaunt that every rib showed through their black and tan hides. Their feet were so sore from running that blood often oozed on the crusted snow and frozen ground. Yet they always wanted to go again, to get back to the mountains. And somehow or other, they knew they were going—and soon, too. So they cried their excitement to their master sitting by the cozy fireplace.

"Ting-a-ling! Ting-a-ling! Ting-a-ling-ling-ling!"

"Neighbor Hobson's ring again," said Mr. Lowe.

Mrs. Lowe frowned. "I've a good notion to plug the bell so no one can call you today," she said.

"Oh no, my dear, let's not do that. Hunting lions is my job and it must be done in winter when the deer are herded together. I'll have to go out today. I can feel it. Maybe it's because the dogs know they're going. You might as well get a few things ready for me. A little lunch, matches, revolver, cartridges—most of all, dry socks. That'll be all I need."

"Wow-o-ow! Wow-woo-ow! Bar-row-ou-u!"

"Did you feed them this morning?" asked George.

"Yes, I gave them a whole pail full of meat scraps. There was enough for a dozen dogs, but those three gobbled all of it in less than a minute. I never saw such greedy hounds."

"They're not much worse than I am," George grinned. "Last night I thought I never would get enough steak, potatoes and gravy. Say, those hot biscuits with butter and honey were good—and so was the apple pie and coffee. And when you gave me bananas with cream this morning—ham, eggs and three stacks of hot cakes dripping with maple syrup—say! I'm hungry again. When are we going to have dinner?"

"I might find some chocolate cake and whipped cream. It won't be dinner time for three hours yet."

"Ting-a-ling!" went the telephone. "Bow-ou-ou!" came from the shed.

"That's our ring this time. I'll answer it." George got stiffly out of his chair.

"Hello—yes, this is George. Oh, hello, Frank. How are you?"

He listened for a moment. "What am I doing? Why, I'm sitting by the fire. Yes—just got in last night. What's that? What did you say? Is that so? Why, yes—of course I'll come. Right away. I'll get somebody to take me up to Pete King ranger station in a car, then I can get to Selway Falls tonight. Tomorrow I'll be at Three Links and on the trail of those kitties. Goodbye, Frank."

He hung up the receiver and turned to his wife. "It was the forest ranger. Lions are killing deer by the dozen on Three Links Creek. Last night two does were killed close to the ranger cabin."

Mrs. Lowe took a deep breath and nodded her head. "Then you'll have to go," she said.

Next evening as George trudged slowly on snowshoes up the Selway river he came upon a freshly killed deer near the trail. The three big dogs which had been following him broke into a frenzy of barking and yowling. They smelled the big cat tracks in the snow and knew it was a fresh trail. George grabbed Darby in his arms. He yelled, "Ike! Ike! Ike!" at the other two to quiet them. It was too late in the evening to start. Better go to the ranger cabin and wait until morning. The big cat would not go very far, and man and dogs were too tired to spend a cold night outside, when a cabin was so near at hand.

Next morning they returned to the slain deer and the dogs took up the trail with a joyous song, a low, happy, "Wow-ou-ou-oo-ou!" The cat tracks were cold. There could not be much of a cat odor to them, but the dogs ran with their noses to the ground, confident that in a short time the cat smell would be plainer. In a minute the dogs were out of sight, but George could hear them. He did not try to hurry. He knew the dogs could easily find the lion and chase it up a tree. George followed the trail because it was easy to travel that way. It was a trail much used by deer. In about a quarter of a mile from the carcass there was another dead deer.

"The old waster!" George muttered. "He wasn't satisfied with killing one deer. He had to kill two, and he never ate one bite out of this one."

By this time the dogs were singing from the top of a low ridge a mile away. All at once their voices took on a new note. There was some excitement now in their cries and their song was louder.

"The trail is hot already!" George exclaimed and took a straight line up the hill, through the pine trees. Right on top of the ridge was a lot of brush where deer had been feeding. The snow was covered with tracks. Now he heard the dogs down the hill two miles away in a cedar swamp. They were singing even more loudly. The mountains rang with their deep, joyful song as they ran faster and faster on the fresh tracks of the mountain lion.

Still George did not hurry. There was plenty of time. He looked around on top of the ridge and found a third dead deer. None of it had been eaten, either.

"Three deer for that old fellow in one night!" he said. "This is the first time I ever found more than two killed by one lion in one night."

Across the canyon Darby, Reno and Custer cried their loudest. They had the lion on the run. The big cat was headed for a bit of rough, rocky ground in the forest, where he could hide. He ran hard, but not hard enough. The fast running dogs, with their terrifying cries, were right behind him.

The mountain lion weighed two hundred pounds. He was six feet long. His teeth were sharp and so were his claws. Would he stop and fight? The three lean dogs all put together would not weigh more than a hundred pounds. They were not fighting dogs. Their job was to trail and bark. Why didn't the lion stand his ground and kill them? He had chased, caught and killed three deer in one night. A few nights before that he had killed a fully grown cow elk. Surely he was strong enough to kill three scrawny, noisy dogs.

The cat has almost everything in its favor. It has the body

of an athlete, lithe, strong, perfect muscles that move with the speed of light. It has a tough body not easily hurt in a fight. The cat is a great fighter. Just get one cornered and you will see!

Is it courage that the cat lacks?

Hardly, when one of them will tackle a fully grown elk. Most mountain men, though, say that a lion is cowardly. They say that if a man happens on one in the forest and scares it badly enough, it will run up a tree. "Just yell as loud as you can, shoot your pistol, throw your hat at it, run straight for the long-tailed rascal and it will take to the nearest tree," says the lion hunter. "Any little dog can tree a big lion. It is the greatest coward on earth!"

The mountain man must be wrong. The cat does not really lack courage, but mental balance. The old mountain lion is easily confused. He knows that a yowling dog, tearing in and out at him, soon has him so that he doesn't know which way to turn. A dog speeds around and around a lion on the ground, first one way, then another. It muddles the lion's slow brain and pretty soon he doesn't know anything. Instead of being cool and collected, a dangerous fighter, a thing to be feared, he is a hopeless, helpless, confused creature that strikes out blindly at anything that moves. So, before he gets cornered, he takes to a tree top, safe from the hounds that make such a frightful noise.

Custer, Reno and Darby did not like cats of any kind. For thousands of years their ancestors had disliked cats. They were of a breed of dogs that had been trained to hunt and sing a song of the chase. It was their life and they loved it. Now they raced through the forest, noses ever closer to the mountain lion tracks, singing their best.

George Lowe heard his dogs, heard the tune that told him the dogs were hot on the trail, heard the other tune that meant the lion was running for a tree. There was silence for just a moment. That meant only one thing. The big cat was up a tree!

"Yoop! Yoop-oop-oop!" Slowly the dogs began to sing. "We have him treed! We have him treed! Treed!" was what they meant.

Mr. Lowe came up on his snowshoes. The dogs were sitting under the tree.

"Yoop! Yoop! Yoop!" they said.

George snapped leashes in their collars and tied them to small bushes, at a safe distance from the old Tom lion who was crouched on a limb. The dogs sulked. They did not like to be tied up. They wanted to be in on the kill. Sometimes Mr. Lowe did not kill a lion with one shot. Then it tumbled down fighting with all its might and main. Dogs that rushed in on a wounded lion were apt to be badly crippled.

With his dogs safe, Mr. Lowe shot the lion. It came crashing down, hit the ground with a thud, and lay perfectly still. The dogs were frantic. They made the woods ring with their barking.

George turned them loose to nip, snip and bite at the dead lion. That was what the dogs worked for, a chance to see the big cat dead and worry its carcass a little. They were satisfied now and lay down in the snow while George skinned the lion.

"It's only ten o'clock in the morning," he said as he looked at his watch and then at the dogs. "Shall we get another one today?"

"Wow! Wow-ow-ow" "Bo-woo-woo-oo!" "Ou-ou-ou-u!" said the dogs, coming to their feet and ready for another chase.

"This way, then," said George and started off toward a long ridge ten miles away, the dogs following his snowshoe trail.

Darby found a porcupine track and stopped to sniff at it.

"Ike! Ike! Ike!" shouted George. "Come back here."

He raised his arm as if to throw something at the dog and Darby left the track. "You come along! Don't you dare chase a porky!"

It was noon before more lion tracks were found. They had been made by a beautiful, sleek female lion. She had four almost fully grown kittens at the head of a gulch under a black granite cliff. Sometimes they killed for themselves, but more often the mother hunted for them. She was a good hunter. As silently as a bird's shadow she crept through the forest. Her padded paws made no sound on the snow. On rocks or frozen ground she pulled back her long claws into furred sheaths so that they made no noise. In brush she bellied low so as not to be seen.

On the brow of the ridge, six deer nibbled at the snowberry bushes. The lioness crouched low in the snow. Nothing showed but her ears and the tip of her long, restless tail. She wormed her way behind a tree only fifty yards from the deer. She lay there for a little while and watched. They were down the hill from her. The way was clear. She waited until all the deer were busy nibbling at the bushes. Then, singling out one of the six deer and gathering all her powerful muscles, she leaped into a burst of speed that took her down to the startled deer before they were aware that there was a lion anywhere near them. She covered the fifty yards in less than four seconds and landed in one last mighty bound on the back of a young buck. He had no time to do more than raise his eyes to the streak of yellow fury that bore down upon him. Both deer and lioness went down in a heap. Her teeth were locked in the deer's neck. Her four clawed legs gripped his body. The lioness had killed her prey.

The big cat feasted. She brought her kittens and they feasted, too. Then came a pack of howling, bawling demons on their trail. George Lowe and his dogs had picked up the lioness' trail where she went to the cliffs for the kittens. The dogs travelled that way, at first. The old lioness, hearing them go away, realized that there would be more time for her to escape. Through a wide stretch of timber, down a steep hill, up another hill, and across an open mountain meadow was a rough, rocky, broken country where the brush grew thick and where there were many caves, many holes between the rocks, places where the dogs would not dare to follow. She and the kittens started on a slow run. They did not try speed. They would become exhausted too soon if they did. Cats can run fast for only fifty yards or so.

As lioness and kittens ran through the timbered land and down the hill, the cry of the dogs grew fainter. Once they were so far away the cats could not hear them at all. They kept right on, though, and started up the hill. They soon heard the dogs again, yelping with excitement as the trail grew warm. The lioness and kits tried to hurry, but going uphill is a poor time for cats to hurry, and the dogs gained on them. Every minute the cries were closer, louder, filled with more danger. The cats strained their muscles. Their lungs were almost bursting when they reached the top of the hill. Behind them the dogs cried their loudest.

George was four miles away. "Dogs mighty close to five cats. I'd better hurry and help them," he said and snowshoed as fast as he could toward the noise.

Halfway across the level, open meadow the dogs overtook the lioness and her cubs. The dogs made a mistake here. They should have waited until the cats got to the first trees, then they would have had them, right where a dog likes to have a cat, up a tree. Instead, Reno darted in at one of the kittens, Darby at another and Custer at a third. The fourth kitten ran. The lioness crouched low and waited. Darby was so excited he did not see her as he circled the kitten with dazzling speed and terrifying yowls. The other dogs were busy, too, each with a kitten. None of them was paying any attention to the lioness.

She did not leap at the tormentors. She did not rush at them at a bear would. All she did was wait. Soon Darby circled too close to her. She reached out a front paw, with bared claws, and struck the dog a glancing blow. He yelped a different tune and scrambled out of the snow a dozen yards away to find out what had happened to him. The lioness still crouched on guard. Darby rushed at her. She reached out and struck him again, squarely on the head this time. Two of the kittens got away to the forest. Reno and Custer treed them in low, scrubby pine trees. A third kitten came along and they put it up a tree, too.

Darby was in bad shape and howled his hurts as loudly as ever he howled a hot trail. Reno and Custer were crying a steady "Yoop! Yoop!" George hurried as fast as he could. He could tell from the way the dogs cried just what was happening. "They're all alive, anyway," he thought as he puffed up the hill.

The lioness took advantage of Darby's hurts to run to a point of timber across the meadow and escape in the rocks.

George took Darby in his arms and carried him to where Reno and Custer had three kittens cornered. He laid him in the snow, tied up the other dogs and shot the young lions with his revolver. Then he built a fire, dressed Darby's wounds the best he could, and skinned the three kittens. By now it was almost dark, and the cabin on Three Links creek was twelve miles away.

Neither man nor dogs had eaten a bite since breakfast. They were hungry and tired. George wanted to hunt down the old lioness and the other kitten. If he did not they would kill many deer before spring came. If he went to the cabin he would have a long, tiring walk in the dark. Darby couldn't walk. He would have to be carried. It is not easy to walk on snowshoes and carry a dog. It would be midnight, perhaps, when he got there. Then, as soon as daylight came, he would have to come back to this spot and start on the trail of the old lioness.

George looked at his dogs. "Guess we'll stay here tonight," he said.

The dogs did not mind. Their life was filled with cold, hunger, and dangerous work. All they asked, or expected, was to be with their master and hunt down the big cats. George managed to shoot two snowshoe rabbits with his revolver. He cooked them over the fire and divided them with the dogs. Then he spent all the rest of the time until dark gathering fuel for the fire. He had no axe, so he had to take such dead limbs as he could break off standing trees.

Night came on, cloudy and cold. The hunter and dogs huddled the long night through by the camp fire. Smoke blinded them. One side of their bodies toasted while the other side froze. This kept them turning around and around. There was no sleep, no comfort, and little rest.

Daylight came at last. It got light enough to travel. They left Darby tied up by the fire while they circled to find the tracks of the lions. It did not take them long. The dogs set up their tuneful cry of cold trail. In an hour they were crying hot trail and in less than another hour George heard the joyful cry, "Yoop! Yoop! Yoop!" and knew he had another mountain lion.

It was the kitten, instead of the old one. It was cornered on top of a big granite rock. George killed it and got back on the old one's trail. It led to a cave far back in the cliffs. It was dark inside the cave. The dogs wanted to go in after the lioness, but George would not let them. "We'll go back to the cabin, wait a day, and then look for this old girl. She won't stay here long."

Darby was well enough to travel in a day or two; and three times in the days that followed, George and the dogs went back to track down the lioness.

She came out at nights and killed deer. She had to. There was nothing else for her to eat. Every time, though, before morning, she went back to the rocks and caves where dogs and man could not force her into the open, or up a tree.

George waited near the cave entrance one whole night. She did not stir from the depths of the cave that night. It was so bitter cold that George's fingers and ears froze, right by the camp fire. When morning came he was so numb he could hardly walk. The dogs moved as if they were numb, too. Stouthearted as they were, they could not keep a whimper out of their voices at the misery they felt.

George thought of his home beside the Clearwater River, the warm room with its blazing fire, a hot bath. More than anything else, perhaps, his thoughts dwelt on a sizzling beefsteak, fried potatoes, coffee and apple pie.

"Say, dogs! Let's go home!"

Reno, Custer and Darby looked up at their master. Their mouths lolled open, and the pleased expression on their faces told how willing they were to do what he wanted them to.

"Besides," George grinned at them, "that old pussy cat has earned her freedom—for another year, at least!"

BEAR FAMILY

(URSIDAE)

Ursidae is the Latin term denoting the bear family.

Bears are the largest in size of the meat-eaters. There are two kinds of bears in Idaho, **grizzlies** and **black bears.** A big grizzly may weigh one thousand pounds.

Nat. Park Service

GRIZZLY

Ursus horribilis is the Latin name for grizzly bear, meaning "horrible-bear." This is a good name, too, because the grizzly is a bad actor when wounded or cornered. Mother grizzlies defend their young with a ferocity unequaled in the whole animal kingdom. If a mother grizzly even thinks her young might be harmed, she flies into a terrible fury and attacks the threatened danger.

"Silver-tip" is another name for the grizzly. The long black hairs have a tip of white which gives them a silvery color. Sometimes a very old bear may be a rather light grey color. Young ones are apt to be nearly black.

Because grizzlies are so destructive to cattle, colts, sheep and pigs they have been hunted down and killed in most of the United States. There are still a few in the forests and mountains of Idaho and other western states. Deer and elk can outrun the lumbering grizzly if they see him coming in time. But these big bears are adept in stealing close to a band of deer and making a swift run at them. If there is a crippled or aged deer in the herd, the bear has a good chance to catch and kill it. One good blow with a mighty front paw is enough to knock a good sized elk to the ground.

A grizzly will kill and eat any other animal, no matter how big nor how small. Sometimes one of them will go to an enormous amount of trouble to dig out a ground squirrel or chipmunk. He likes ants, honey, grub worms from rotten logs, berries and fish. In fact, a grizzly will eat almost anything that any other animal will eat, including a small amount of grass and weeds.

These bears are very intelligent, but because they are so dangerous and destructive it is not wise to allow them to live close to people. A stock grower would be foolish indeed to allow a mean old silver-tip to roam around his ranch killing cattle and sheep at will. Silver-tips have been known to kill people, too.

So the place for this shaggy old fellow is a long way back in the mountains where he can cause but little trouble and damage.

Black bears have the Greek name of *Euarctis* which is "true-bear." This is one of the most familiar of all our wild animals and is found in the mountains and forests of nearly all of the rougher lands in the United States.

Black bears are naturally happy-go-lucky, playful fellows. They are good natured and inoffensive. A black bear is very intelligent and can be taught to do many tricks. However, they play too rough to be good pets.

When first born in February, a cub is hardly as large as a kitten, but grows very fast. By mid-summer one will weigh as much as twenty pounds and in two years grow to two hundred pounds. They sleep the winter away, as do all bears in a cold, snowy climate.

I no see him, but I can tell. He break my shirt, now he kill my sheep!"

Once Franz saw a black bear kill a sheep and he fired a shot. The black bear ran with a bad limp as if it might have been hurt in the shoulder.

You don't suppose it was Geolu with caked mud on his yellow coat, do you?

GEOLU'S WINTER SLEEP

While huckleberries were ripe, Geolu and the other bears ate little else. They liked elderberries, too, and broke down the bushes to get them. Sometimes there would be five or six bears in the same berry patch. They paid little attention to one another. Sometimes one would get too close to a mother bear and her cubs. This always drove the mother into a fury and she charged with wide open mouth, to rip and tear, unless the intruder beat a hasty retreat. Mother bears allow no familiarity with their babies, even by another bear.

After the berries were all gone and fall came on with a snowstorm, the bears had a hard time finding enough to eat. All the sheep were taken to winter ranges. Ground squirrels and marmots went into their holes deep in the ground, even before fall weather began. A bear often worked several hours to dig one ground squirrel out of its den. Later on, the soil froze hard and the bear could not dig through to get even one small ground squirrel to eat. An occasional mouse, chipmunk, or pine squirrel was about all a bear could find. The food situation was critical.

Geolu had fed well on the sheep he had killed. He had eaten greedily of the luscious berries and was very fat when the first snow fell. Then he wandered far into the mountains. One day he heard rifle shots. Some were far away, some near by. He had to be very careful now, or an elk hunter would see him and kill a bear instead of an elk. So he kept to dense growths of ceanothus, laurel, young groves of pine trees, willows or other hiding places where hunters would not be apt to find him. Geolu did not travel around very much and make lots of tracks in the snow for hunters to see, either.

Then one day he smelled freshly-killed meat. He did not go near it, even if he was hungry. It might be a trap and wise bears do not blunder easily into traps. He waited until there were no more rifle shots before investigating. This meat he smelled was an elk that the hunters had killed. Fortunately for Geolu, the hunters had left most of the elk meat. He feasted on this for several days, then wandered about the mountains in ever-deepening snow.

There was not a bite of food anywhere for Geolu. Yet he had enough food stored up to last him from October to March. Sounds queer that he had a store of food, yet not a bite to eat, yet that's the way things were with Geolu.

Geolu came to a deep canyon with dark, forested slopes. He was sleepy and decided to find a good place and go to sleep in this canyon. There was a choice of the north slope or the south slope.

Why in the world should there be a difference between the north slope and the south slope, when it came time for a bear to go to sleep?

There was quite a difference, as Geolu knew very well, a difference of about three weeks' sleep. He was very fat, so he decided on the north slope and the three weeks' extra sleep.

You see, Geolu meant to sleep through from fall to spring. Spring came fully three weeks earlier on the south slope, even though it was only a mile from the north slope. The sun's rays struck warmly and the southwest winds blew gently on the south slope to melt the snow away. The north slope was in the shade and protected from melting winds, so winter lingered there.

Geolu found a great pile of broken branches and brush where a giant fir tree had fallen to the ground. Beside the trunk of this big tree and under a heap of debris, he dug out a den. He did not make it large and roomy, but only big enough for him to crawl inside and lie down comfortably. He pulled some brush into the opening behind him to seal the entrance, closed his eyes and fell asleep.

Snow came down—and more snow until the mountains were covered, four, five, six feet deep. Winds blew out of the north. It got cold, ten below zero, twenty below and once or twice forty below zero.

Geolu slept the time away, warm and cozy. He hardly breathed. The blood barely circulated in his veins and arteries. His stomach was empty when he went to sleep, so there was no digestive trouble.

Hold on there! How about that store of food Geolu had for the winter! *Where* is it and *when* does he use it?

The food supply is fat stored up under that fluffy yellow hide. Slight as was his breathing and blood circulation, some energy was required to keep them going. Little by little, the fat was consumed. Food that had been laid away for the long sleep was gradually used.

In February, there was a warm spell, a chinook. It even rained hard all one day and the whole country was slush and water. Geolu must have stirred uneasily in his sleep during this time. He might have thought that spring had come. But he didn't. He stayed right where he was with the snow frozen hard over him and slept for another month.

The south slope was pretty and green when warm winds and rains finally melted the snow on the north slope. One day Geolu was rudely awakened. Water was trickling down through the brush and debris over his den, right into his face. He was fully aware that spring had come. There could be no doubt about it this time. He pulled the brush away and crawled out of his den.

The hair on one side of his body was matted. There was a rubbed spot on one hip where he had lain against a stone. Geolu had slept in one position for nearly five months. He was a little stiff from inactivity. His feet were very tender and as he walked over the rough ground, they became so badly cut and scratched that he left a trail of blood wherever he went. He walked slowly to the south slope and began to look for food. There were old rotten logs to tear apart for white grub worms. Certain plants were dug up for their bulbous roots, but even so, Geolu became very hungry after a few days. All of his winter fat was used up, his long yellow hair loosened and began to drop out. The nights were chilly. Geolu thought of the juicy sheep he had eaten the summer before and started to travel that way. At a small stream he stopped to fish. He loved to stand in a creek and get a meal of delicious trout which were on the way to their spawning grounds. Fish was a main article of diet for a few weeks. Again he started toward the sheep ranges. It was June now, and the sheepherders were on their way to the mountains with their bleating charges.

One morning Geolu came down to a level stretch of land beside a large creek. Here were some fine ant hills. He liked to drag a front paw across an ant hill, watch the insects scurry hither and thither, then lick his long red tongue into the thickest of them and grind them between his teeth. Ants were a spice, a sauce and relish in the bear's diet. This particular morning, he found another bear at his favorite ant hill. He stopped at the edge of the forest to watch it scratch the ant hill and lick up the tasty morsels. It was a pretty black bear with a sleek, shiny coat, a bit smaller than Geolu. He turned once to go away and seek another ant hill. For some reason, though, a reason that Geolu did not quite understand at the time, he stopped to look at the black bear again.

That moment changed the whole course of Geolu's life. That was the moment that decided the question of whether he would go out to the sheep ranges, kill sheep, and sooner or later be killed by a sheepherder like Franz, or whether he would live a long life as a law-abiding bear, far back in the mountains.

For when he looked, the black bear looked back at him. Slowly he walked forward, his new silky coat of yellow hair fairly glistening in the sunlight. The other bear waited. Geolu walked up to her and they rubbed noses. Geolu had found a mate. She led him away with her, not toward the sheep ranges, but far back into the mountains.

RACCOON FAMILY
(PROCYONIDAE)

Raccoons have the Greek name of *Procyonidae,* meaning in English "before-dog-family," and referring to their position in the animal kingdom.

These are rather scarce in Idaho. Sometimes a trapper gets one in a trap set for coyotes. They are rather pretty animals with a long ringed tail.

A very beautiful near relative of the raccoon is the **cacomistle,** or **ring-tail.** Often desert miners and prospectors have one of these for a pet. They are about the size of a small house cat and have a very long black and white ringed tail. Ring-tails are good mouse catchers and keep a cabin free from these pests.

Idaho is just a little too far north for these warm desert loving animals, but occasionally one is seen in the southern part of the state. Probably it has been brought to Idaho by some traveller or miner.

Ring-tail

WEASEL FAMILY

(MUSTELIDAE)

Mustelidae is the Latin name for the weasel family. These are small meat-eaters, compared to the bears, but unequaled in ferocity.

This family includes the **weasel, mink, marten, fisher, wolverine, skunk, otter** and **badger.**

BLACK-TIP, THE KILLER

Black-tip lay in the snow with only his eyes showing. He lay in wait for his quarry which was coming across the snow in wide jumps. Straight toward Black-tip the huge white animal bounded. It was ten times as big as Black-tip and ten times as strong, yet he was not afraid. Often when hungry, Black-tip had killed prey as large as this one that was now almost upon him.

BLACK-TIP THE KILLER

At just the right instant Black-tip leaped. He grasped his prey with his four legs and sank his sharp canine teeth deeply into the back of its neck. It uttered a cry of fright and leaped wildly to one side in an effort to dislodge Black-tip. It weakened after a dozen jumps. Black-tip tightened his grip. He sank his fangs in a new place. There was a crunching of teeth against spine and the big animal rolled over dead. Then the weasel feasted upon the snowshoe rabbit's muscle and blood.

Black-tip dived into the loose snow, as a beaver would dive into water. There was only a small round hole to show where he went under. Then he travelled on the ground under the snow until he found a warm place beside a log where he could lie and rest. There were plenty of mice under the snow. He could kill some of them when he grew hungry again. He often killed chipmunks, squirrels and blue grouse, too, even when he did not need fresh meat to eat. Sometimes when he killed he took only some of the animal's blood and left the rest of its body lying in the snow.

Black-tip loved to come out on top of the snow at night when the moon was shining. Then he made tracks through the snow. He could make so many tracks in one night that a person walking through would think there were a dozen or more weasels in that vicinity. Up and down the slopes, in and out of the dense thickets, under and over fallen trees he circled, doubling back and crisscrossing his trails. His footprints were tiny, about like a person would make with two fingers, one held a little ahead of the other. A foot or eighteen inches farther on there was another print, and another. Hind feet came down almost exactly in the prints made by front feet. Black-tip's tracks were neater and more precise than those of a squirrel or rabbit. These show all four foot prints. The hind feet are spread wide and make a separate print in the snow from the front feet. In loose snow the squirrel, or rabbit, makes a sort of blotch. The weasel always makes a trim, even footprint.

The start of Black-tip's bewildering maze of tracks is a small, round hole where he emerges from the snow. The end is another round hole where he goes under for the day's sleep and rest. Although he prefers night time for cruising and killing, the weasel is often seen in broad daylight.

Once Black-tip came in his wanderings to a deep, wide, smooth trail in the snow, running straight through the forest. The trail smelled good. It smelled of meat. This was something new. No weasel can resist investigating new things, so Black-tip followed to see where it went. After awhile the trail turned to a big spruce tree. There was a deep notch cut in the tree about three feet from the top of the snow and a small pole was leaned into this notch. The meat smell was strong on the pole. A piece of meat was fastened in the notch on the opposite side from the sloping pole. What in the world could all of this mean? Black-tip was not hungry. He never ate anything except freshly killed meat, anyway, but he just had to run up the pole and investigate the notch and meat. Inside the notch was a steel trap set the day before by a trapper who had snowshoed through the country. The trapper thought he might catch a marten in the spruce swamp. He had dragged a chunk of meat behind him in the trail so that any animal getting the scent would follow along to his steel trap set in the side of a tree. The trapper had even rubbed the meat on the leaning pole so it would smell, too.

Now Black-tip had never seen a steel trap and did not know of the danger that threatened him. One of his tiny front feet on the trigger of that trap would snap it shut like a vise on his front leg. His slim, supple white body would struggle and he would fight back with all his marvelous strength, but the steel trap would hang on. It was different from the flesh and blood antagonists that the weasel fought with.

Black-tip sniffed at the jaws of the trap, he even sniffed at the trigger. The blood smell was strong. He walked across one jaw of the trap to sniff at the meat fastened in the opposite end of the notch. He backed up against a steel jaw, turned and put his front feet across to the other jaw, spanning the trigger with his body. It was close quarters in the notch for even a small animal like the weasel, but Black-tip managed to turn around and run back down the pole. He didn't know that he had been flirting with a death much worse than any he had ever seen, even for a killer.

THE WEASEL

It is a weasel's nature to kill. It is its way of living. A rabbit lives on green leaves and never kills anything except, perhaps, plants. A squirrel eats nuts. A beaver loves the bark and wood of aspen and willows. Mink eat fish. Each and every animal must live according to the way nature has fashioned it. We cannot blame a weasel any more for being the most ferocious killer in the whole animal world than we can blame a rabbit for eating the tender leaves of plants in the forest.

The weasel does a lot of good in keeping down mice, ground squirrels and other crop pests. As long as it confines itself to mice and other rodents we like it. We say the weasel is a beneficial animal. When it gets out of bounds and raids the chicken-house, killing a dozen pullets in one night, we get pretty angry at it and set some traps. Just like we lose our tempers when wild rabbits come to the garden at night and nibble our lettuce plants.

We have said that grizzly bears are the largest of the MEAT-EATERS. The weasel is the smallest.

The **Rocky Mountain weasel** is only six inches long, including its black-tipped tail. Inch for inch, ounce for ounce, this little fellow is the champion fighter of the animal kingdom.

In the summer time these little meat-eaters are a light brown color, except for the black tip of the tail. When winter comes, they turn pure white, all except the black tail tip.

Their fur is short but very, very soft. It is sold as "ermine." A lady's short jacket made of first grade weasel skins and ornamented with weasel tails may sell for as much as five thousand dollars. The white skins are also used to trim ladies' coats and dresses.

There are weasels all over North America from the Gulf states to Alaska. One species in Idaho and farther west has a body length of eleven inches and a tail seven inches long—eighteen inches total length, three times as long as the Rocky Mountain species.

Martens, Latin *Martes,* are the most beautiful of all the meat-eaters, unless it is a silver or black fox. They are about the same size as a small house cat. Their fur is a rich brown color and very soft to the touch. Martens have large erect ears and big, night-seeing eyes. A long, bushy tail adds to their beauty. They live in the dense forests and high mountains, feeding on blue grouse, pine squirrels, mice and other small mammals and birds. Because of their valuable fur, martens must be protected from over-trapping. If not, they would soon be exterminated.

Martens become quite tame around mountain cabins when not molested. Once one of them came into a cabin late one evening where a forest ranger was reading a book. The marten came through the open back door, advanced into the light of the gasoline lantern and looked the ranger over curiously as if to say, "Who are you and what are you doing here?"

The ranger sat very still. He wanted to know what the marten was going to do just as much as the marten wanted to know what the ranger was going to do. The marten came nearer, reared to a half sitting position, and looked at the ranger for a half-minute. Then he ran over to a pail of water sitting on an up-ended apple box. There was plenty of water outside, in fact, a good-sized creek ran past the back door of the cabin. Just the same, the marten jumped up beside the water pail, put his fore feet on the rim and took a drink of water from it. Then he ran back to the ranger, stopped to regard him keenly, and seemed to say, "Thanks for the drink, pal. So long." Anyway, the marten turned and ran out of the back door well satisfied with his visit to the ranger's cabin.

The ranger was glad he had not frightened the pretty animal. "Maybe it will come back again some night and call on me," said the ranger to himself.

A **fisher** is a big marten. It, too, lives in the forest and mountains. A fisher's body is two feet long, tail fifteen inches. An average fisher weighs about twelve pounds. It has a valuable fur and one of them is a prize for any trapper. They are very scarce.

Mink are about the same size as martens. They have small ears and small, wicked-looking eyes. Mink look more like

MINK

weasels than martens do. They live on frogs, fish, snakes, muskrats and birds.

Mink live along the streams from Mexico to Canada. Their fur is dark brown and black colored, very soft and silky, and very expensive.

These animals can be successfully reared in captivity. There are a great many mink farms in the United States that raise mink for their fur. Northern Idaho has some fine mink farms.

Wolverine. This is a big, sturdy, long-haired, crafty weasel, often called the "glutton" because of its greedy feeding habits. A wolverine is about thirty-three inches long, has an eight inch tail and weighs thirty pounds. Its long fur is not very valuable, nor does it have many enemies, but it is quite rare, just the same. A few remain in the most remote parts of the mountains.

The glutton has a very bad reputation for breaking into mountain cabins, destroying trap lines set for martens or foxes, and killing animals it does not need for food. It has inches and *ears* nearly *one and one-half inches long*. Imagine an animal with ears nearly as long as its body. In front of the ear is a projection called the tragus. You can easily find the tragus of your own ear. Just put your fingers to your ears. The little flap next to your eye is the tragus. It seems to act as a sort of sound magnifier. In a bat the tragus is very large and shaped to give effect to the tiniest sound.

With such ears the most sensitive vibrations are *heard* in time for the bat to twist away from whatever makes them.

Another kind has the family name of *Phyllostomidae*, Greek for "leaf-nosed." This form of bat is not native to Idaho, but is extremely interesting because of the intricate folds of skin around the nose and mouth which give its face the appearance of an outlandish flower. These folds of skin are extremely sensitive to air currents such as are encountered in flying toward an obstruction. The bat *feels* in time to change its course.

The lump-nosed bat has a large and homely nose. It probably is a very sensitive one. What could be more likely than that this highly developed nose *smells* out dangerous obstructions in the line of flight.

As proof that a bat does not need to see with its eyes, experiments have been made with the bat's eyes fastened shut. It flies just as swiftly and safely without eyes as with them.

Bats have scent glands which give off a nasty odor that other animals abhor. In its way, the bat is almost as bad as the skunk. No other animal, unless it is the owl, can endure this disagreeable smell, much less enjoy eating the animal that produces it. Its tough wing membranes and long ears in which it wraps itself to sleep are not very good to eat, either. As an article of diet for other animls, the bat rates very low.

They sleep in daytime and go forth in the evening to find food. Bats hang by their hind feet, head down in sleeping. With a full stomach of insects they come to their roost, make a half roll as they grasp whatever twig, rock or bit of bark they choose to hang from, and go to sleep. Some species congregate in large colonies. Sometimes they hang in clusters, like a bunch of grapes, to sleep. Other species live singly or in small colonies. Hollow trees, crevices in rocky cliffs and hillsides, attics, barns, sheds, undersides of bridges, abandoned mine shafts and tunnels, as well as natural caves are the homes of bats. We are apt to find them in cities as well as the country.

Hanging head downward, a bat is in perfect position to take flight again if molested. It is much safer from attack than a bird in its nest, a mouse in its hole, or even a rabbit in its burrow.

Bats sleep most of the time. Some of them sleep the winter away. It is quite probable that a bat sleeps on the average about eighteen hours a day.

Bats have thin, squeaky voices. At least, that is all a mere person can hear. With their exceptional ears, it is quite probable that they have a world of sounds whose vibrations do not register on our crude ears.

In so many ways bats are the most remarkable of all the mammals. There are many families and several hundred species in the order of bats. In Idaho there is one family (Latin name *Vespertilionidae)* and twelve species. There are seven species of *Myotis,* which is Latin for "mouse-ears." Their ears are plenty big, but small compared with other species. These are small bats with bodies about two inches long, tail a little more than an inch and a half, wing spread about ten inches. Four species are known as **little brown bats,** the other three being called: **golden long-eared, interior long-legged** and **black-nosed.** The five species besides the mouse-eared ones are: **silver-haired, Western canyon, big brown, hoary** and **lump-nosed.** Their names describe them quite well.

The bodies of all Idaho bats are covered with a fine fur, like that of a mouse. Even part of the wing is furred. A bat's wing is not at all like a bird's. The bat has a hand with four extremely long finger bones. The forearm bones are very long, too. Between the fingers and from them to the body, hind legs and tail, there is a very fine, light, tough membrane. When it is extended until taut it makes a most efficient wing. It really is a winged-hand, and the Greek name, *Chiroptera,* is descriptive. The fifth finger or thumb, is a crooked claw, useful in climbing and fighting.

Bats live on insects which they gather on the wing out of the air. All species of Idaho bats are harmless. They are even beneficial to some extent in keeping down the supply of harmful insects.

The bad reputation of bats is due to some of the species found in other parts of the world. The vampire bat, found mostly in South America, is a horrid creature. It is a naked-skinned, very ugly, repulsive little beast, four inches long, which does considerable damage to horses, cattle and other animals by biting and sucking blood. Its bites are deep, and blood flows freely. This weakens and finally exhausts its prey. The vampire bat has even been known to bite people.

Another notorious kind of bat is the "flying fox" of New Zealand, Philippine Islands and other parts of the Old World. The flying foxes are fruit-eating animals and do a great deal of damage to orchards. Some of them have a wingspread of four or five feet and as they gather in flocks of hundreds of thousands, they darken the sky, and their wing beats make a noise that can be heard for miles.

Idaho bats are very small in size compared with the enormous flying foxes. The Western canyon bat, or *Pipistrellus*, is one of the smallest. It has a body length of less than two inches, and wingspread of about eight inches. The most familiar bat in Idaho is probably the little brown, which is not much larger than the Western canyon bat.

We have heard of the wonderful things that a bat can do, of its superiority over other animals in many ways. How about its faults? What are its weaknesses?

In several ways bats are the most frail of all animals. They cannot walk. They cannot even raise their bodies on all fours, much less stand erect on their woefully weak hind legs. They can crawl along after a fashion, and some species get part of their food by crawling on the ground for bugs. Their strength, however, is in their wings. In any other situation except when flying, the bat is all but helpless.

How about those delicate wings? Does anything ever happen to them?

There! You have hit on the weakest spot in this highly developed mammal. An injury to a wing, even a slight one, is fatal to the bat. A bird can fly minus a few feathers. They will grow back in due time. Then too, a bird is not wholly at a loss on the ground. Not so the bat. A slit in the wing membrane, or a broken finger bone, means death.

Well, you may say, there isn't much danger of a wing's

getting injured. That is true enough. The bat has almost everything in its favor. It hides away in the daytime. It has an odor that makes it repulsive to other animals. The mother bat carries her youngster along wherever she goes, so no harm may come to it while she is feeding. Bats sleep upside down, ready for instant flight. Of all the mammals, their senses of hearing, smell and touch are most keen. They are the most agile in flight and feed on a supply of insects which is always abundant. When insects are not at hand, the bat either goes where there are plenty of them or goes to sleep and waits for them to come back. There are no enemies. Why do bats not overrun the earth?

Sadly enough, the bats are their own enemies. They are quarrelsome, greedy and selfish. No other animal on earth, except *man,* is so quarrelsome with its own kind. The bats use their canine teeth, sharp as a weasel's, and their wicked thumb claws to tear at one another's delicate wings. They take one another's lives. If it were not for this fatal weakness, there might be black clouds of bats over the sky every evening.

ORDER OF GNAWERS

(RODENTIA)

SQUIRREL FAMILY	(Sciuridae)
POCKET GOPHER FAMILY	(Geomyidae)
RAT AND MOUSE FAMILIES	
Pocket Rats and Pocket Mice	(Heteromyidae)
Native Rats and Mice	(Cricetidae)
Old World Rats and Mice	(Muridae)
Jumping Mice	(Zapodidae)
BEAVER FAMILY	(Castoridae)
PORCUPINE FAMILY	(Erethizontidae)

Rodentia is the Latin term denoting order of gnawers.

Rodents have chisel-like front teeth, two above and two below, for cutting and gnawing. These incisors grow out from the roots as fast as they are worn away on the edges. Rodents have no canine teeth. Their back teeth (molars) are adapted to grinding food such as seeds, nuts, bark of trees, shrubs and other vegetable food. Most species have, as part of the intestines, a sac called the caecum, almost as large as the stomach itself, for the storage of partly digested, coarse, fibrous food.

In North America there are more than eight hundred kinds of rodents, out of a total of over fourteen hundred kinds for all the orders of mammals.

There is a great difference in habits and modes of living in this order. Some squirrels live in trees and are wonderful climbers. A few species of tree squirrels can glide such long distances between trees they are called "flying" squirrels. There are many species of rodents that live underground and are good diggers. Beavers and muskrats build dams and houses in the water and are great swimmers. Kangaroo rats inhabit the driest desert.

Some species of rodents can be found in almost any part of the earth from sea coasts to the highest mountains.

In this order there are also great differences in size between the various species. The smallest rodent is also about the smallest of all mammals. It is a pocket mouse found in desert country and has a body less than two inches long. The largest rodents in Idaho are the beaver and porcupine which may weigh forty pounds or more and measure forty to forty-four inches from tip of nose to tip of tail. The largest rodent on earth is the capybara of South America, which weighs about one hundred forty pounds.

Rodent fur is of many kinds, from the soft silky fur of some mice to the heavy durable fur of muskrat and beaver. Porcupines have a coat of long greenish yellow hair in which needle sharp quills are liberally scattered.

Some rodents such as brown rats, beavers, muskrats and many species of mice have hard, scaled, naked tails. Others have tails covered with fine down. A few have gorgeous bushy tails. Perhaps you have noticed the beautiful tail of Neotoma, the wood rat, or those of the tree squirrels and woodchucks. And of course you know about the porcupine's tail, covered with horrid quills.

Most species of rodents multiply very rapidly if not kept in check. They soon overrun their food supply and die of disease or starvation. Meat-eating birds and meat-eating mammals depend largely upon rodents for food. People are inclined to kill all owls, hawks, weasels, coyotes and snakes because they fear these meat-eaters will take poultry or sheep. Then people have to wage war on rodents, too, or they would take everything.

In cultivated fields, orchards, gardens, granaries, dwellings and irrigation ditches, many species of rodents must be controlled by shooting, trapping or poison. Wherever people live and cultivate the soil they make conditions good for mice, rats, ground squirrels and pocket gophers. Cultivated soil is easier for them to dig in, and the crops raised furnish them with an easy food supply.

Where the soil is not cultivated, ground digging rodents do considerable good. They dig up the soil so that more rain water soaks in. They carry to their dens and runways vegetable matter which rots and enriches the soil. Their tunnels allow air to come in contact with the soil and that is beneficial.

There are several quite serious human diseases which rodents are guilty of spreading in one way or another. Spotted fever is carried from rodents to people by wood ticks. Sylvatic plague is carried by a short-hopping flea. Tularemia is carried by insects or transmitted direct to people handling small rodents, either dead or alive.

It is a good practice to look out for wood ticks during the spring and summer. Always scrub the hands thoroughly with soap and hot water after handling wood ticks or skinning rabbits. Never handle ground squirrels unless it is absolutely necessary. Then always wash and disinfect hands thoroughly afterward.

THE SQUIRREL FAMILY

(SCIURIDAE)

Sciuridae is the Latin word denoting the squirrel family.
There are seven important groups in this family.

Group One is the **woodchucks.** If you wish to be more technical, call them marmots. Some people incorrectly call them ground-hogs or hedge-hogs. Other names are: whistling pig, whistlers and rock-chucks. The Latin name is *Marmota* which means "marmot."

This is the animal, so the old story goes, that comes out of its winter den on February second and forecasts the weather. If he sees his shadow on that day it will be bleak, wintry weather for six weeks longer, so he goes back to his den to sleep some more. If he does not see his shadow, the weather will be good and and he stays out! Why not keep track of February second for a few years and find out how good Mr. Marmot is as a weather prophet?

These rodents live in ground burrows, usually around big rocks, but sometimes in fence rows and old log piles. They feed on vegetation of many kinds and sleep away most of the winter. Marmots give a loud, shrill whistle when alarmed.

There are four species in Idaho: **British Columbia woodchuck,** in the extreme northern part of the state: **yellow-bellied,** in the western portion; **golden-mantled,** in the Rocky Mountain region; and the **Montana hoary marmot** of the Bitterroot and Salmon Mountains. The British Columbia woodchuck is smallest in size with a body length of about fifteen inches and a tail five or six inches. It is very reddish in color. The golden-mantled is the most beautifully colored, yellowish and reddish with a golden buff mantle across the shoulders. The Montana hoary marmot is a big fellow and almost white in color. Its body measures twenty-one inches and its tail nine inches.

Woodchuck meat is greatly esteemed in some localities. Indians are very fond of it. The animal's fur is not of very great value, but leather made from the skins is very fine, soft and strong. It is desirable as lace leather in fancy leather are apt to become pests if they get inside on the table or in the cupboards.

Group Seven is the **flying squirrel**. *Glaucomys* is the Greek name, meaning "blue-grey-mouse," which is a poor name for it does not describe the flying ability of this remarkable animal.

The flying squirrel is a glider rather than a flyer. It has no wings to beat and cannot go upward or sideways. It simply extends its feet and legs out as far as possible to stretch the loose skin of its belly into a broad flat surface and glides from a high point in one tree to a lower point in another tree twenty, thirty or maybe forty feet away. Then it must climb up like any other squirrel if it wishes to make another long glide.

There are two species in Idaho, twenty-three in the United States.

The Bangs flying squirrel ranges in the mountains of central and eastern Idaho. It is six to seven inches long with a tail about five inches. Its belly is pinkish in color, its feet grey.

Broadfoot is larger with an eight inch body, a six inch tail and a cream white belly. Its range is the mountains of northern Idaho.

The flying squirrel is the only member of the squirrel family which prefers night to day. Rarely is one seen in the daytime, unless in a trap, or in its nest in a hollow tree. Often they venture into mountain cabins and stare curiously at the fellows beside the stove or table in the light of tallow candles or kerosene lamps. Their fur is very silky like a young rabbit's; it is not valuable because of its poor wearing qualities. Flying squirrels can be easily tamed and make good pets. They never learn, however, to like broad daylight.

Marten trappers do not like flying squirrels because these gliders get into the marten traps before a marten does. The trap is sprung and the catch is worthless to the trapper.

Food of flying squirrels consists of nuts, grain, insects, birds and berries. Meat is eaten when it can be secured. Their worst enemy is probably the owl. Weasels and martens are also bad enemies.

Not a great deal is known of the habits of the flying squirrel because it is such a difficult animal to study. Night-active animals do not tell us as much of their life stories as day-active ones do.

THE POCKET-GOPHER FAMILY

(GEOMYIDAE)

Geomyidae is a Greek word denoting "earth-mouse-family."

Thomomys is the Greek name for the **Western pocket gophers** and means "heap-mouse." This is because of the gopher's habit of making a heap or pile of dirt in building its tunnel under the surface of the ground.

Pocket gophers are digging rodents with external fur-lined cheek pouches. They have small eyes and ears, strong digging claws, and rather short, very sensitive, scantily haired tails which serve as a guide when the gopher retreats backward in its tunnel. They are good-sized animals—a big old one that lives in loose soil may measure all of eight inches long with an additional four inches of tail.

These rodents are seldom seen as they spend almost their whole existence underground. They are clever tunnel diggers. The digging is done with the fore claws and teeth. The dirt taken from the tunnel is heaped out on the surface every few feet through short upright tunnels. Sometimes the gopher finds food on the surface of the ground near the mouth of his tunnel, but he never ventures very far outside. Always he plugs up the entrances with dirt to keep out light, which he shuns. Plugged tunnels also prevent snakes and other animals from using his feeding grounds and home.

The gopher is strictly a vegetarian. It eats bulbs, tubers, roots and other underground growths, also surface food such as alfalfa, grain and garden vegetables. It is often very destructive to crops and even more so to orchards; it is particularly fond of the roots of apple, pear and other fruit trees.

The pouches in its cheeks are used to store and carry food. A gopher caught near the mouth of a tunnel nearly always has his cheek pouches crammed with food. This shows that he gathers surface food as fast as he can so as not to be outside the safety of his tunnel any longer than necessary.

It is easy to capture a gopher if you can find an open hole. Step carefully so as not to jar the ground too much and scare the gopher away. Put a loop of strong cord around the hole, stand back about ten feet with the end of the cord in your hand and wait for him to stick his head out. Jerk the loop tight around his neck. You will have a fighting animal on your hands and one that will give you a nasty bite if he gets a chance.

Ordinary steel traps are almost useless as the gopher will bury them with dirt. Special traps are better, but if gophers must be destroyed the most effective way seems to be poison. The natural enemies of pocket gophers are hawks, owls, snakes, weasels, skunks, coyotes, foxes, badgers and bob-cats. Owls and badgers are probably the worst enemies. An old owl can afford to sit on a fence post for a long time waiting for a gopher to stick its head out of a hole, to swoop silently and swiftly down to grasp it in wicked talons. A badger, of course, being a good digger, captures gophers underground. Gophers are very prolific and increase rapidly where soil and food conditions are good.

There are a great many species. Some of them are so nearly alike that it is difficult to make clear distinction. The **Pigmy pocket gopher** of southeastern Idaho is only a little over four inches long with a tail two inches long. It is a rich brown in color. The **Idaho** is a bit larger, pale yellowish grey with very small ears. The **Townsend** of lower Snake River Valley has a body length of over eight inches with tail of four inches, making it a little over a foot long from tip to tip. This one is dark grey or black in color.

On rough, uncultivated lands these ground tunnellers do a great amount of good. Their tunnels take up surface water which soaks into the ground instead of running off to muddy up the rivers. They stir up the soil by carrying some of the sub-soil to the surface and they enrich the soil by carrying vegetable matter underground to decay.

When it comes to farms and orchards—oh, my! They must be thinned out!

RAT AND MOUSE FAMILIES

First of all, let's take a look at the mouse most common to everyone. It is, of course, the **house mouse**. Its family name is *Muridae,* which is Latin for mouse. It is very dull colored dusky brown, black and ashy grey. Its tail is partly naked. While its home is usually in houses and barns, it is often found in the fields with other mice.

This house mouse is an undesirable alien and a nuisance wherever it lives. It is destructive, ill-mannered and an ugly little pest.

Then, there is the **brown rat** which belongs in the same family. This is the fellow which gives the whole rat family a bad name. It is the black sheep of the family, seen around garbage dumps, in old buildings, grain bins, barns and even sewers. Any place where there is easy food and drink you are apt to find the brown rat. It will not live where it has to make an effort to find food. On farms, brown rats kill little chickens, eat eggs and gnaw into granaries. The slums of great cities are alive with rats. Sometimes they become so bold they come right into rooms where people are sitting, in broad daylight. Rats have been known to attack people.

Brown rats are infested with fleas which carry diseases to people. This is another reason we think of rats with horror. The very word has a distasteful meaning. One definition of "rat" in the dictionary is "a person who deserts his friends in hard times." So, all in all, we do not like rats, and are glad to know that the brown rat is not a native American animal. It came to this country on ships and has multiplied until there are now millions of them almost everywhere in the United States.

The brown rat is a muddy brown color. It is seven or eight inches in body length with a naked, round tail about the same length. Its ears are large and naked, too. Now we have said some pretty strong things about the brown rat. We might sum it all up by calling him a disgrace to the animal kingdom, a fellow of which the other animals might well be ashamed.

The hand of man is justifiably against the brown rat wherever it is to be found.

How about the white rat? The pretty, pink-eyed fellow that makes such a good pet?

You may think it strange, but this white rat is nothing more nor less than a brown rat with white hair. It is a domesticated brown rat and has all the bad qualities of its brown brother which is allowed to run loose.

The **pocket rats** and **pocket mice** have cheek pouches like the pocket gophers. The Greek name of this family is *Heteromyidae,* meaning in English "different-from-other-mice-and-rats-family."

The main difference between it and others is the cheek pouches in which it carries food.

Pocket mice have the Greek name of *Perognathus,* meaning "pouched jaw."

These are really beautiful little creatures and it is a pleasure to know them. There are five different races of them in Idaho. The principal differences in these races is color and size. A very pretty one is the **Oregon pocket mouse,** body a little over three inches with a tail nearly four inches long. It is grey or buff colored. The largest one is the **Idaho.** Its body is about three and a half inches and its tail about four inches long; its color is light buff mixed with black. Other kinds are **Great Basin, Uinta** and **Northwest pocket mice.** Another species still is the **Pacific,** smallest of all mammals. This one has a body only a bit over two inches long and its tail is about two inches. It is found near the Mexican boundary.

Pocket mice live in the deserts and plains of the state, usually where the soil is loose and digging is easy. They make tunnels and deep burrows in the ground but, unlike the pocket gophers, these mice come out on the surface at night and run around much as other mice do. One never comes out of its underground home in the daytime. Their presence on the desert is made known by their burrows and runways on the sand. Look for their tiny tracks in the sand next time you are in a dry country!

The cheek pouches are used for carrying and storing seeds. A supply is stored up in their burrows for winter use.

Pocket mice are easily tamed and make fine pets. Don't try to get them to drink water, though. A pocket mouse—pocket rat, too, for that matter—doesn't know that water is good to drink. They can live for months with only the moisture that is in their food.

Pocket rats are also called kangaroo rats. *Dipodomys* is the Greek name, meaning "two-footed mouse." Of course they have four feet, but the hind legs are two or three times as long as their front ones, and in running the front legs are not used. They jump along, very much like a kangaroo, on their hind legs.

Kangaroo rats live in the prairies, arid plains and deserts. A kangaroo rat is the exact opposite of its distant cousin, the brown rat, in every way.

It lives cleanly, in a house of its own building, usually in or near a clump of brush at the base of a sandhill. This house is built of sand, gravel and small sticks, two or three feet high and six to eight feet in diameter. Not the least bit of a sneak, like the brown rat, the kangaroo rat builds its house in the open for all to see. Inside the house is a network of tunnels and burrows. Rattlesnakes and lizards are guests sometimes, probably very unwelcome ones, but the kangaroo rat makes no visible attempt to chase them away. Trails radiate from the house in every direction across the sandy wastes. They have been made by the rat on his trips for food. He eats whatever plant life he can find. He lives where there is no water and if captured and offered water to drink he will refuse. He has no way of knowing it is good to drink.

Kangaroo rats are among the most beautiful of all mammals. Their bodies are about five inches long, covered with soft, silky long fur. Black, white, grey, tawny brown and cinnamon buff spots, stripes and patches give them a pretty color pattern. Their bellies and feet are pure white. They have long tails, all of seven inches, marked by four blackish-and-white stripes, a white ring at the base and a brown tuft of hairs at the tip. They have large eyes, big round ears and fur-lined cheek pouches for carrying food.

In disposition this desert rat is admirable, too. He can be trusted not to bite while being handled, unless he is hurt or

MUSKRAT

siderable damage to irrigation ditches and levees by tunneling through the banks. Small streams of water trickle through these small tunnels and quickly become large enough to make the bank or levee give way completely.

Besides plants, muskrats feed on such animal life as fish, mussels, frogs and salamanders. Their most relentless enemy is the otter, which can dive under water, find the entrances to the muskrats' home, go inside and kill. Other enemies are hawks, owls, mink and weasels.

There is one species of muskrat in Idaho along the streams, in lakes and swamps over most of the State.

One more family and we are through with rats and mice. This is the family of **Jumping Mice.** *Zapodidae* is the Greek name. In English this means "very-(long)-footed-family" and was given because of the extraordinarily long hind legs. These mice are the jumpers of the rat and mouse clan. Two feet is a good high jump for Zap, and three or four feet for a broad jump. When running to escape an enemy he goes like a kangaroo rat, in a series of high, long jumps without using his front feet. Zap travels a zig-zag course in order to baffle his pursuers.

Should you find a long-legged, exceptionally long-tailed, yellow-colored rodent in a meadow, brush patch, forest or other rather moist place, it will be Zap. The only other animal anywhere near like him in looks is the kangaroo rat and it has cheek pouches and is always found in dry places.

Zap's long tail acts as a balancer and rudder in jumping. Deprived of this he would make mighty awkward leaps. Zap takes on fat during summer and fall for his long winter sleep.

There are four species in Idaho, differing from one another slightly in color, size and mode of living. They cover most of the State except the main Snake River Valley from about Weiser to Idaho Falls; one species lives south of Snake River to the Idaho-Nevada-Utah boundary.

BEAVER

(CASTORIDAE)

The Latin name for the beaver family is *Castoridae*. These are one of the most common of all western mammals. They have webbed hind feet for swimming, hand-like front feet for handling building material, a double claw on each hind

BEAVER, CHEWING GREEN ASPEN

foot for combing out their fur, and very large, curved front teeth. Beavers have played an important part in the exploration and settlement of the West. They still are important as conservators of water and soil, besides being a source of very fine furs.

BEAVER RANCH

Dan and Warren Hilton trotted their saddle ponies from the barn across the small pasture to a wire gate a half mile away. Dan dropped off his horse, opened the gate, laid it to one side so that a small band of sheep could wander out into the large pasture. He remounted and the two boys galloped the horses across the big pasture to the west gate, went through it, turned to skirt a field of growing wheat and came to the creek

DAM THE BEAVERS BUILT

that flowed from the mountains. Back toward the Hilton ranch they rode at a walk. It was eary morning, before breakfast, yet the sun was already an hour high. Spring had come. School was out. There was plenty of work to do on the ranch, but they could snatch an hour every morning to ride the creek and see how the beaver were progressing with their dams, canals and timber cutting. The beavers on the Hilton ranch were a serious problem.

This was the fourth season the two boys had watched the beaver. The first year there was just a pair of them, two miles upstream. They made a small dam which backed water up over an acre or two of pasture land. Mr. Hilton was interested in the beaver then and went with the boys several times to watch their progress. No one could guess where the beaver came from. There were none closer than Swan Lake, ten miles away across a ridge. Evidently this pair had traveled that distance to their new home on the Hilton ranch.

They had selected a spot near a grove of aspen trees. Dan and Warren found it while they were looking for a stray horse.

"Look, someone is chopping down our quaking aspens," cried Dan as they rode upon the scene.

"Some woodchopper!" said Warren, gazing at the gnawed stumps of the small trees.

"It's beavers," Dan decided.

The boys dismounted, counted a dozen stumps, found the trail where sections of the tree had been dragged and followed it down to the beginnings of a beaver dam.

The beavers knew exactly what they were about. No engineer with surveying instruments could have done a better job in locating a place for a dam across the creek. Ten feet farther down was a rapids. A dam at this spot would need to be quite high. It would be difficult to start a dam, too, in swift water. Farther up from the place picked out, a dam would have to be longer and the lake behind it would be smaller.

Warren said, "They sure are smart. Notice how they have a downhill drag from where their cuttings are, to their dam. No uphill pulls for them."

"Yes, and a little later when water begins to back up behind their dam, they can float their logs in," Dan said.

"Logs, sticks, stones and mud. Wonder how they handle those rocks?"

"Let's sneak up on them sometime and watch them work," suggested Dan.

Thus began a long series of trips to the beaver colony for the two boys. They watched the beaver dam grow from its beginning, with only a puddle of water behind it, to a strong, four foot high structure and a good-sized lake. They watched the beaver house grow and the grove of aspens dwindle. It took an enormous number of trees and great quantities of stones, sticks and mud.

The beaver usually started work in late evening. Warren and Dan never did find out where they hid in the daytime, although they boys hunted for them several times. Always before starting to work, the beavers swam around their pond and looked things over. The first piece of timber added to their house or dam at the start of a night's work was ridiculously small. Sometimes it was only a twig, not much longer than a lead pencil with, perhaps, a few leaves on it. Later in the night, they tugged with all their strength at chunks of aspen tree, four or five feet long and five or six inches in diameter. Some of the smaller parts of the trees were cut longer and pushed downstream across the top of the dam as braces against the pressure of the water. Mud was carried under the chin, held in place by the forefeet. The beavers' front feet were much like hands, and they used them like hands, too. The larger stones were pushed or rolled in place. Few stones were used, but plenty of mud was carried to cement the sticks and logs together. It was a strong dam.

The house was built on a circular floor plan. Mud, sticks and stones were carried for the house just as for the dam. A canal was started from the center of the house, and this was the only entrance. Every night the two beavers worked on their dam, home and canal. Their whole enterprise was carried on without one part being neglected for another. And it was perfectly designed. The beaver knew exactly where the level of water would be in their house. Above this, they arranged their living quarters, a wide semi-circular floor covered with shredded wood. To go outside, they had but to slip off the floor into the water, dive down into the canal, swim under water for thirty feet and come up in the center of their lake.

The beaver cut down trees at night, too. Dan and Warren watched them one night. The two beavers did not set upon a tree, one on each side and chew away at it. They seldom worked on the same tree at the same time. Nor did they work constantly on one tree until it fell. One would chew for awhile on one tree, standing on his hind legs, braced with broad tail and gripping the tree with his hands. Then tiring, he would go to another tree. Sometimes the two worked on five or six aspen trees, one after the other. They made some good-sized chips with their long, curved, chisel-pointed teeth. The boys learned that beaver could chew on the roots of trees, or gnaw at the bark of logs under water, without ever getting a drop of water in their mouths.

How?

Their lips are large and loose. Under water, the beaver opens his mouth wide to gnaw, but closes his lips behind his cutting teeth. His nostrils and ears close, too. He can stay under a few minutes very comfortably. He must come to the surface for air, or drown, even if he is an expert swimmer and diver.

When fall came and it got cold, ice froze on the beaver pond and sealed them in. Snow came and covered their tepee-shaped stick and mud house. They could not go out for food. This bothered the beaver not one bit. They had stored up a lot of aspen wood on the bottom of their lake for winter food. All they need do was dive into the canal, swim along under the ice to the food supply and satisfy their hunger.

At first, Dan and Warren were mystified when they saw so much aspen cuttings on the bottom of the lake. They thought a beaver must possess magic to get the wood to sink. It should float, so the boys thought, until Dan cut a length of green aspen and tossed it into the water. It sank. There was no mystery to it. The wood was simply heavier than water.

Next the two boys found out that beavers do not always get their trees to fall toward the water. They marked some trees and later on saw these same trees felled by beaver. They lay the way they leaned. Some did not even fall, but lodged in other trees and hung there all winter long. The animals were plenty smart, but they made some mistakes.

Next spring, there was no home to build, so the beavers played more. Shortly after the ice went away, Dan and Warren saw three baby beavers playing with the two old ones. They dived, swam and splashed water with their broad tails. Sometimes when one of the old ones slapped the water, all dived out of sight. It was part of the game they were playing.

Old Lutra, the otter, came the next winter during the Christmas holidays, when ice was over all the pond except a small spot near the dam where water spilled over. He had come overland at night from Swan Lake, ten miles away. There was snow on the ground and his tracks showed plainly. His long, muscular, short-haired tail dragged in the snow and made a track like a big snake might make. No one could mistake Lutra's trail. In summer he had fed on fish, but now most of his favorite fishing streams and lakes were frozen tight. He found a muskrat family, killed and ate all of them. Now he was after beaver. He came upstream over the snow and ice and climbed up the beaver dam until he could look over. His web-footed tracks showed where he had walked up and down on the dam. He saw the beaver house.

Lutra's mouth watered and his beady eyes glittered. Next to fish, he liked fresh beaver meat more than anything else, and he was hungry now, too. How to get at their stronghold was a problem. If one of them should happen to come out on the ice, it would be a simple matter for the fast-moving otter to catch the rather slow-moving beaver. It was unlikely, though, that one would leave the pond. The beaver had no need to go out from under the ice to feed, for there was a big supply of food in a safe place.

Lutra walked over the ice and snow to the beaver house to listen. He heard the beaver move about. He climbed on their house and sniffed the warm air that came from inside, through an opening in the top. Then he went back and slid into the water to search for the canal that led to the entrance of the beavers' home.

All this Dan and Warren could read in the snow. The tracks were plain.

They read more. There were other tracks leading away from the beaver pond and there were red stains along this trail where Lutra's blood and been spilled. The track made by the long tail was not as nicely curved as it was coming over from Swan Lake. Lutra seemed to be trying to hold his tail to one side. Most of the blood stains were from his tail. Evidently he was trying to keep from dragging a painful wound against the snow. Searching closer, the boys saw that a front leg had been dragging, too. So they pieced the whole story together.

"Lutra found the canal and got inside the house, all right," said Warren.

"He got a warm reception, too. Our beavers were not caught napping," said Dan.

"There are no signs he killed any of them."

"If he had got one, he would have stayed here until he killed them all."

"I'd like to have seen him poke his old weasel head up in the house and get slashed on the shoulder by old man beaver."

"And while one was slashing at his shoulder, another took a chip out of his tail."

"Served him right."

"We will tell Mr. Hawkes and he will set a trap for that old killer," Dan finished.

In the spring, another pair of beaver came from somewhere and settled downstream a half-mile. Dan and Warren watched them build their dam, house and canal. The third year, still another colony started. There were plenty of young beavers and they mated, built homes and cut aspen trees, too.

"We will have to do something about all those beavers," said Mr. Hilton. "They are ruining the creek. Why, some of them have even dug out houses in the banks and have underground tunnels to their lakes. Soon there will be no trees left anywhere. They come out in the fields to eat alfalfa and wheat. I can see their trails a quarter of a mile from the creek."

"We hate to have them killed," said Dan, and Warren agreed. "We like to watch them work."

"I'll let them go one more year," said Mr. Hilton. "Then something will have to be done, if they keep on increasing."

Now Dan and Warren were riding this early morning along the margin of what had once been a beautiful aspen grove bordering the creek. It was desolate looking, indeed. Unsightly stumps, trunks of larger trees and brush were all that was left of the once lovely woodland. Dams and beaver houses were plentiful.

"Awful looking place, isn't it?" Warren remarked.

"Who would have thought all this would come from one pair of beavers four years ago?" asked Dan.

"There must be twenty or thirty of them now."

"Wonder what Dad will do?"

"I hope he lets some of them stay."

They were almost within sight of the ranch house when

Dan's horse broke through the ground over a beaver tunnel and plunged down with both front feet, throwing Dan headlong from the saddle. There was no real harm done to either horse or rider, but Mr. Hilton saw the accident. He had seen more, too. There were fresh beaver gnawings on some apple trees in the orchard.

"Those beaver must be cleaned out at once!" was his decision.

Dan and Warren looked at each other. That was what they had feared.

"I'll get a permit from the Game Department to trap all of them."

"Couldn't we leave at least one family at the upper end of the creek?" asked Warren.

"Well, maybe," said Mr. Hilton without much enthusiasm. "I'll tell you what I'll do. You fellows do the trapping and skinning and I'll give you half the money you get for the hides. They are still prime."

Dan and Warren looked at each other again. This was a tempting offer, twenty to thirty skins, at maybe ten dollars apiece. It would be a sizeable amount of money, yet it seemed a shame to kill any of those interesting animals.

"Of course, if you don't want the job, I can get old man Hanker and his steel traps," said Mr. Hilton.

"We'll do it," Warren agreed suddenly, accepting for his brother, too. "We'll go into town this morning and look up the Conservation Officer."

"O.K.," said Mr. Hilton and went about his morning duties.

Dan and Warren were all excitement when they came back from town. They would answer no questions. "You just wait until tomorrow—then you'll see!" was all they would say about their plans.

The other folks at the ranch caught some of the air of mystery from the two boys before the day was over.

"I wonder what they're up to?" Mrs. Hilton said.

"Oh, nothing," the boys' father answered. "It's just the thought of making so much money that has them all keyed up."

"No, it's more than that. They're acting just like they did the time they found old Bess in the pasture with a white colt and kept it secret until they could bring the old mare and colt down to the barn for us to see. Those boys are up to something." Mother Hilton knew her sons.

Next morning a truck drove into the yard and a Conservation Officer got out. He had a half dozen of the strangest beaver traps anyone ever saw in the back of his truck. They were long cage-like affairs made of springs and wire netting, designed to catch beaver alive. He also had lumber, wire netting and posts to build a small pen.

"Let's put up the holding pen first," said Ben Hale, the officer.

BEAVER IN LIVE TRAP

"We've got a place picked out," said Dan, and led the way to a spot behind the spring-house.

"Fine," said Ben Hale and peeled off his coat to go to work.

They made a small square pen under the trees, lined it with woven wire and put some boards across the top for a roof. Inside they put a couple of old wash tubs and filled them with water. The boys brought armloads of willows from a place a little way down the spring creek and put them in the floor of the pen. About noon the officer pronounced the pen satisfactory.

"It's O.K., fellows," he said. "We'll set traps this afternoon."

They began on the beaver colony nearest the orchard. There was a fresh beaver trail, still wet with water that had dripped off the beaver's fur, leading out of the pond. One of the big traps was set with wide, meshed sides gaping open, under water in just the right place for a beaver to swim against the trigger when he started for the trail. Farther up another trap was set, and so on, until six of them were out.

"Now we must wait. Just before dark we will take a look," said Ben Hale.

Three beaver were caught that evening, all young ones. They were carried to the holding pen, trap and all, then released on the bed of willows. The traps were taken back and set in the same places. Next morning there were five beavers and one big carp in the traps. The trigger of one had been set so lightly that a carp swimming against it had released the meshed sides and caught the fish.

"I don't get your idea," said Mr. Hilton. "Why not catch them in steel traps, kill them and skin them outright on the spot? Why go to all this fuss?"

"We are not counting on killing them at all," said Dan.

"What?" Mr. Hilton was astounded. "What in the world are you going to do with them?"

"Ben Hale is going to haul them to the mountains sixty miles from here and turn them loose."

"What for, I'd like to know?"

"Up there they will do no harm to anyone, and they will do a lot of good," said Dan.

"Yes," Warren put in, "Their dams will help prevent floods."

"And hold the soil from washing away."

"And make places for fish."

"And when their ponds silt up and they move away, they leave a patch of good rich soil, where grass will grow for deer, or sheep to eat."

"They are worth five or ten dollars for their hides, but up in the mountains they are worth a hundred dollars, maybe, for the dams they build."

"Goodness sake!" said their bewildered father, grabbing his hat and hastening toward the barn.

"Are you boys going to the mountains with your beaver?" asked their mother.

"We'll have to stay here until all of them are caught," Dan began.

"Some of them will have to go to the mountains today," put in Warren.

"And some tomorrow. Our pen isn't big enough to hold all of them."

"And they fight something terrible when they are crowded."

"They go to different places in the mountains, too."

"There is a lot of work to planting beaver."

"We want to go with the last bunch, way up in Buffalo Mountains."

"Come out and look at the ones in the pen."

Mrs. Hilton, as bewildered as her husband, walked out to the pen and watched the boys pet and handle their beaver. Dan reached down, grasped one of the smaller ones by the broad tail and lifted it outside the pen. It did not offer to bite or struggle. Dan stroked its back, but was careful to keep his hands a safe distance from its sharp teeth. After a bit, he lifted it back to the pen.

"See how easy they are to handle?" he said.

"When you know how," Warren grinned.

"What I want to know is how you two have found out so much about beavers," said Mrs. Hilton.

"We haven't been watching them three years for nothing," Dan laughed.

"I've a notion to go to the mountains with you and see where you put them," said their mother.

Mr. Hilton came up just in time to hear the conversation. "I'm going along, too," he said.

"Fine!" the boys shouted. "Fine!"

They caught twenty-four beaver altogether and twenty of them were hauled away by Ben Hale. The remaining four were an old pair and two young ones. These were taken from the pen one fine spring morning and started for their new home in the mountains.

"How many do you think are left on our place?" asked Mr. Hilton as they drove down the road away from the ranch.

"Not more than one colony, with maybe a few strays."

"Yes, we left one or two scattered ones."

"How many years before we have all this to do over?" Mr. Hilton frowned.

"Well, if we take them every two years, there won't be so many."

"About a dozen beavers every two years."

"They won't get as far down as the orchard."

"No, they'll stay on the upper end of the creek."

"Huh! Guess we may as well call our place the Beaver Ranch!" Mr. Hilton tried to be stern, but his growl lacked fierceness.

That afternoon Ben Hale found the place he wanted. It was a small patch of willows on both sides of a tiny creek. Aspen trees were plentiful. Ben drove his truck right up to the spot and the Hilton family followed on foot, leaving their car on the road a half mile away.

Ben and the boys unloaded a box-like affair about three feet wide and four feet long, made of rough boards, and carried it down to the creek. They dug out a hole in the bank of the creek with shovels until there was water in the bottom. They placed the box over this and shoveled mud against it.

"A good beaver house, I'd say," said Ben Hale. "Bring the beavers, boys."

He removed a loose board on top of the box and put the four beavers through the hole. There were shelves inside the box, far enough above the water so the beavers could lie on them. A hole in the side under water would allow the beaver to go outside.

Ben and the boys shoveled mud on top. The boys found some old logs they could handle and with some boards from the truck they made a dam across the creek.

That night at the Hilton home, Mr. Hilton leaned back in his chair and looked sternly at his two sons.

"My bargain with you called for me getting half the returns from the sale of those beaver skins. Where do I come in on this deal? I have no beavers and no money."

Dan and Warren looked at each other with worried faces.

They had thought about this very thing. They had let their enthusiasm carry them away. After all, Dad owned the ranch and had offered to give them only half the returns from the beaver trapped. True enough, he had said, "Trap the beaver, I'll give you half the money you get for their hides."

There had been no money from the trapping, so there was none to divide. They had tricked their father because he never meant to give the beaver away.

They could not answer. They hung their heads.

"Do you think those beavers will get along all right, way up there in the mountains?" Their father tried to make his voice harsh, but they could detect softness in it. "There are otters, coyotes, bob-cats and all kinds of wild animals up there that might eat them up. Poor beavers!"

"Oh, they'll do all right!" said Dan.

"They'll take care of themselves," Warren added.

"Well, I'm going back to see—and if they're not, I know two boys who are going to work pretty hard to pay their father half the value of twenty-four beaver pelts."

Three days later the Hilton family drove to the mountains again. Dan and Warren hung back with their parents when they left the car, instead of racing ahead to look at the new beaver home. What if, after all, something had happened to them? A coyote might have killed them. They might have wandered away. Homesickness might have driven them back toward the Hilton ranch. A lot of things might have happened.

At the foot of the last rise before reaching the willow patch, the place where the truck had had such a hard time getting up, the boys could constrain themselves no longer. They raced ahead. What they saw brought hearty shouts, "Hurry, Dad! Hurry, Mom!"

"They're here!"

"They're all right!"

Mr. and Mrs. Hilton looked down upon the scene. Already the four beaver had made their plans and laid the foundation for their future home. They had cut a few trees and dragged them in to start their dam. They had started a new house. A scratch over the ground showed where a canal would be. They had even carried a few small branches and placed them on top of the box house Ben Hale had made for them.

It was evening, and as the Hiltons watched, a beaver came down from the aspens carrying a ridiculously small aspen branch, the first work of the day, to place in the dam.

"Well, I'll let you two fellows off this time," said Mr. Hilton, trying to be gruff. "I should give you some kind of a stiff fine, though, for not being better dam locaters."

Dan and Warren looked at each other. Neither spoke.

Mr. Hilton grinned. "Those beavers ignored the dam you built for them and picked out a much better site twenty feet away. Come on, boys, let's go back to the Beaver Ranch!"

PORCUPINE

(ERETHIZONTIDAE)

Erethizontidae is the Greek family name for porcupine. It means in English "irritation-girdle-family," or the family which wears a girdle of irritation. This is a good descriptive name, for if ever an animal was capable of inflicting irritation on another, it is the porcupine.

Porcupine quills are loosely attached to its hide so that the slightest touch causes them to drop off. The quills are two to four inches long, stiff, very sharp and barbed. They enter into the flesh of another animal easily and because of the barbs, continue to work deeper. When it comes to pulling out a quill deeply embedded in a horse's nose or a dog's mouth, it is not so easily done. The barbs pull out wide and make a firm grip.

The porcupine's tail is a deadly weapon. With it, he can strike another animal such a hard blow that dozens of quills are driven deeply into the flesh. Nearly all wild animals let porcupines strictly alone. Only when very, very hungry will a bob-cat, coyote or fisher tackle a porcupine. There are no deadly quills on a porcupine's face and nose, nor any on its belly. An attacker tries to grasp the porky by the nose and turn it over. Then the belly can be ripped open without danger from quills. Since the porky can turn end for end and use its tail in no time at all, any attacker must be lightning quick. Bob-cats and lynx are said to kill porcupines by catching them in deep, loose snow and flipping them upside down with a paw under their bellies. Then they can safely rip and tear at the non-quilled underside.

The porcupine needs a good defense, such as his girdle of irritation, for he is very slow on the ground, has quite poor eyesight and his hearing is not very good, either. He is a good tree climber and spends much of his time in the tops of trees, feeding on the bark. He has an odd craving for salt and will go to almost any length to get a taste. Around old corrals and salt boxes porcupines even eat all the wood which has been in contact with salt. Any tool handle that has dried sweat on it is avidly gnawed. Leather, such as harness or saddles, is greedily eaten.

A person, lost in the woods and hungry, can readily find and kill a porcupine with a club. Then if he is a good scout and can build a fire, he can cook a meal of porcupine meat. If not the best meat in the world, at least it is substantial and that means a lot to a hungry man.

One of the worst things about a porcupine is that in the mountains a person may stumble over one in the dark and become badly hurt.

Porcupines do not do much harm to forests. There may be some places where they kill too many trees, but such places are rare. In thickly settled communities they can do much harm to young cattle, horses and dogs. They may also harm orchard trees.

Porcupines often come right into towns. This is a stupid thing for them to do, as they are always killed.

THE DEER FAMILY

(CERVIDAE)

Cervidae is the Latin term denoting the "deer-family."

All males of the deer family have antlers, and these are one of the most remarkable things about the deer family. Antlers set the deer family apart from all other animals. The main difference between antlers and horns is that antlers are shed every year and a new set grown, while horns are permanent appendages, like claws and hoofs.

The deer family has antlers and the cattle and sheep family have horns.

Just why moose, caribou, elk, white-tailed and black-tailed deer should shed their antlers every year and grow new ones, is a mystery. There seems no good reason for it. As a means of defense against other animals the antler is not very effective, because the deer lose them in late winter, at the very time of year they most need to defend themselves against their chief enemies, wolves and coyotes. Besides, the female deer need protection more than the males, yet females have no antlers at all. (Caribou are the exception here. Both male and female caribou have antlers.) Certainly deer do not need antlers to fight one another. They would be just as well matched without them.

The best explanation is that antlers are an adornment for the male animal, a little something extra that Nature has provided, like bright colored feathers on a bird, a long mane on a lion, or whiskers on a man.

The time of year when antlers are shed differs widely with the species of deer and the locality. Mule deer usually shed their antlers in January and February. A few may shed before Christmas. White-tailed deer shed their antlers later, elk lose theirs in March and moose in April. Caribou, on the other hand, shed antlers very early in the winter.

For an example of how antlers grow, an elk is chosen. The process in other members of the deer family is about the same. Elk shed both antlers early in March and only a stump remains. This stump, or "pedicle," is a part of the frontal bone between eye and ear, and its rough surface extends about two inches through the skin. A new antler begins to grow right away. In just a few days the pedicle is covered with a very tender skin which is well supplied with veins and arteries. In three weeks there is a good-sized tomato shaped bulge on each pedicle. This gradually lengthens into a club-like antler eighteen inches long, with two or more tines branching off by the end of the sixth week. In three months it is full length, perhaps sixty inches, and fully tined. (With a moose we say fully "palmated," or "palmed.")

The soft skin filled with blood vessels is called the "velvet." Inside it, there is a very spongy growth of bone which hardens as it grows. This core is not exactly the same as bone, but nearly so. Bone is composed largely of calcium and phosphorus. Hardened antler is about one-half calcium-phosphorus. The other half is protein and moisture.

Calcium and phosphorus are taken from the forage the elk eats; they are deposited by the blood in the velvet, along with carbon, nitrogen, oxygen and hydrogen, to build up the hard, strong, wicked-looking head ornaments of the big bull elk. Starting with the first tender skin across the pedicle in March; calcium, phosphorus, protein and moisture are carried bit by bit and deposited in the growing antler, much as the sap of a plant carries material from the ground for it to grow.

While in the velvet, the antler is very sensitive and the deer takes great care in going through the forest not to strike it against a tree. If, by accident, an antler is injured even slightly, while in the velvet, it will be deformed. Salt seems to be very necessary to the welfare of all animals that subsist on vegetation and deer will go to rather extreme lengths to get it. Deer and elk often eat out deep holes in clay banks for salt. The males in velvet sometimes get their heads fast in these holes and injure the soft antlers. Often an antler is broken completely off. This makes a deformed, or freak growth. Sometimes a wound, such as a broken leg, or a gun-shot, will cause a deer to have deformed antlers.

If an antler is sawed off close to the pedicle, as is always done when an elk is shipped by truck or rail to some distant range, the short stub drops off the pedicle just the same as a full antler would, when the time comes to shed.

When antlers are fully grown, in July, the skin dies and loosens. For a few days it hangs in ragged shreds like the bark on a dead tree. The big bulls do a lot of "bag punching" when their antlers first harden, using trees instead of regular punching bags. They go through all the motions of a regular fight with another bull—rush, side step, guard, attack, feint, and push the little tree until it is stripped of branches and bark. They keep up this practice for two or three weeks.

In September the antlers are polished white on the tips, very hard and sharp. They are now so firmly attached to the skull that the skull itself will break and give way before one of the antlers can be broken at the joint on the pedicle where growth started.

In their fights with one another, along in September and October, the bulls exert all their great strength in pushing with their antlers. If one gains an advantage he follows it up with wicked jabs that may kill the weaker bull. Often one of the tines will break.

Elk and moose antlers never become locked together in fighting, as those of smaller deer sometimes do, particularly mule deer. The antlers of this species branch in pairs, the tine of each pair branches to another pair, thus forming a "rack" which may readily be interlocked with a similar rack so firmly that separation is difficult or impossible. Two deer thus locked together always die. Their struggles make them weak; exhaustion and starvation finish them.

The antlers are firmly attached until the very day shedding takes place. In some miraculous way the joint loosens and the antler simply drops to the ground. Sometimes both drop at the same instant, but more often there are a few hours between the time when one antler is shed until the other falls, and it is not at all unusual for a bull to carry one antler for a week or more after the other is gone.

Young deer of all species have only small sets of antlers. Elk have "spikes" after they are a year old. Mule deer, and white-tailed deer have tiny forked horns the second year. Moose have small palms. Each year afterward their antlers are larger and have more tines, or points. A mature bull elk, five or six years old, should have six point antlers. A good set should measure about fifty inches on the inside curve of each antler, have a spread of forty-five inches and measure about three inches in diameter at the pedicle.

Deer do not add a tine to their antlers for each year of age. A five year old mule deer may have a dozen points. Size of antlers depends upon the physical condition of the animal, which, in turn, depends upon the quality and quantity of the food. Plenty of the right kind of forage means big antlers. The forage must, of course, be rich in calcium, phosphorus and protein.

Shed antlers on the range are eaten by mice, porcupines and by deer and elk themselves, probably to get the calcium and phosphorus. This is one reason ranges are not cluttered with shed antlers.

MOOSE

The greatest deer of all is the American **moose**. It is so big that the word, "moose," has come to mean any animal, or thing, of extra large size. A big moose is nine feet long from the tip of his ugly Roman nose to the end of his short, ridiculous looking tail. He stands six and a half feet high at the shoulders, nearly a foot higher than his hind quarters. Under his throat hangs a loose fold of skin called the "bell," sometimes more than a foot long.

The eyes, ears and nose of the moose are not nearly as keen as those in other deer, nor is it a fast runner. The moose must depend largely on its size for protection. An extra big moose may weigh as much as fifteen hundred pounds, and the spread of its great palmated antlers is all of six feet. Truly it is a great deer.

Moose are still rather plentiful in certain parts of Wyoming, Montana and Idaho, as well as in Maine, Minnesota, Canada and Alaska. They live in the more swampy lands of the great forests, from the vicinity of Yellowstone Park north into the Arctic Circle. In Idaho they are found on Snake River near Yellowstone Park and in the northern part of the state.

Big game hunters prize the moose as a trophy above all other North American animals. During the mating season in September and October, hunters imitate the "call" of a cow moose, to lure the bulls close enough to kill them. They are not numerous enough in Idaho to allow any to be killed by hunters.

A moose cannot stand on level ground and eat grass. Its neck is not long enough. Most of the moose's diet is shrubs such as willows, aspen, buffalo berry and maple. Often a moose will straddle a small tree, push it to the ground and methodically strip it of bark and small branches. When the moose gets off the tree it straightens up again, and we see a tree twelve, fifteen, perhaps eighteen feet tall, eaten bare of bark and twigs.

It seems impossible to keep moose in public parks for people to admire, as we do buffalo, deer of many species and elk. Close confinement causes moose to die in a short time. They thrive best away from civilization.

Next to moose in size is the American **elk.** Its antlers are round instead of flat like a moose's. A big bull elk stands five feet tall at the shoulders, is about eight feet long and weighs close to a thousand pounds. A big set of antlers has six tines on each side, measures sixty inches following the curves and has a "spread" of fifty-six inches.

When colonists first came to North America there were many elk right down to the Atlantic Coast. Elk lived in western Pennsylvania for many years after white people settled the

ELK

eastern part of the state. Elk meat is about the finest of all wild meat and the animals are quite easily hunted. A good hunter can creep as close to an elk as he wants to, so wherever the country was settled, elk were killed for meat until they were all gone. Now there are big herds in only a few places of the West.

Elk are fine animals for zoological parks. Nearly every zoo in the country has a few of them. From Yellowstone Park and vicinity, elk have been captured and transported to a great many of their former ranges. Advantage is taken of the elks' hunger, in winter when deep snow covers all the grass, and it is easy to capture them. Strong corrals are built and hay is placed inside. The elk seem to know it is a trap and at first won't go near. After a day or two of bitter hunger they venture into the corrals for hay. The gate to the corral is fixed so that a man hidden quite a distance away can pull a wire and close the gate, locking the elk inside. After that, they are handled much like cattle for shipment. They are driven through a chute, up an incline into big enclosed trucks, hauled to a railway station and unloaded into express cars. At their destination, which may be Virginia, Arizona or North Carolina, they are put into trucks, hauled to the forests and released. Elk have been shipped in this way to nearly all the forty-eight states. Some of these plantings have increased to good-sized herds.

There are three different species of elk in the West. By far the most numerous are the **Rocky Mountain** variety. There are big herds of these in Idaho. The biggest ones are on the Selway and Lochsa Rivers in the northern part of the State.

Next in numerical importance is the **Olympic,** or **Roosevelt elk,** which lives west of the Cascade Mountains. This elk is as large as the Rocky Mountain species, darker in color and has shorter and more stocky antlers.

In California is the **dwarf,** or **tule elk.** It is smaller in size, paler in color and has very pale-colored ears. Few of these small elk are left. One band lives near Bakersfield and another in Owens Valley.

Where elk have plenty of food and are not harassed by too many men, dogs, bears, wolves, or unless they are otherwise hindered, they multiply very rapidly. Because they eat so much forage and do not hesitate to raid fields and pastures, elk are apt to become great nuisances near settled communities. When a herd gets too large for its pasturage, the Game Department allows hunters to kill some of the animals in it.

In some of the other states a few years ago when a herd got too large, the shooting season was opened. All hunters that wanted to kill elk were permitted to do so. In some cases there were so many hunters that all the elk in the herd were killed. This was not a very sensible thing to do. It is a good thing to have elk, as long as they do not do too much damage.

A very good system was adopted by the officials of the Idaho Fish and Game Department. They knew about how many elk there were in each herd. They knew how many there ought to be. It was easy to say how many should be killed to keep the herd to proper size, so that damage would not be done to pastures or fields.

Then, instead of letting all the hunters go out to shoot a few elk, only a few hunters were allowed to go. If, for instance, there were a hundred elk to be killed out of a herd of a thousand, a hundred hunters were allowed to shoot one elk apiece. Thus, only a hundred were killed and there were nine hundred left. By the next year, when the little calves were born, there were a thousand again. A hundred more could be taken by hunters. Every year a hundred elk could be taken from this herd of a thousand and there will always be a herd.

Drawings are conducted to determine which license holders can take part in the hunt.

Unless rules like this are made and observed, there will be no wildlife at all in our country in a few years, except, perhaps, some rather undesirable species like rats, mice, English sparrows and insects.

BABY ELK

The **caribou** is the deer of the Northland. The females, as we have said, have antlers as well as the males. The antlers of caribou are not round like elk and the smaller deer, nor are they broadly palmated like the moose, but come in between the two. They are partly round and partly flat.

There are four types of caribou: **woodland, mountain, barren ground** and **Dawson**, or Queen Charlotte Island, caribou. Their home is in Canada and Alaska, although a few range in the northern part of Idaho.

Mountain caribou are the largest, being almost as big as elk. Next in size is the woodland. The other two are smaller. A male barren ground caribou is about six feet long, measures less than four feet at the shoulder, and weighs about three hundred pounds.

The **reindeer** of Alaska are Old World caribou, imported in 1892 from Siberia as a food supply for the Eskimos.

Caribou are rather stolid animals, so easily hunted that it is a wonder any remain to this day. However, there are still great herds of them in the wilds of Canada and Alaska.

There are two groups of smaller deer in the United States, the **white-tailed** group and the **black-tailed** group.

The home of the white-tail is from the Atlantic to the

Cascades in Washington and Oregon. Along the Columbia River they go even farther west, practically to the Pacific. Northward, the white-tail ranges as far as Alberta, southward to Mexico. There are no wild white-tailed deer in Colorado, Utah, Nevada or California. In Idaho their range is the northern half of the State; there they grow to a very large size.

In Idaho a white-tailed deer may be seventy-two to seventy-eight inches from the tip of its nose to the base of its tail. The white, black-tipped tail is a good twelve inches long. A big white-tail stands about forty-two inches high at the shoulder, and may weigh close to three hundred pounds before dressing. The average weight, however, is not more than a hundred and fifty pounds dressed. An antler consists of one main curved stem from which several short tines grow at almost right angles.

WHITE-TAILED DEER

The white-tail has learned a lot about living close to people. Don't think for a moment that wild animals do not learn. If they did not learn new ways of getting food, finding shelter, avoiding enemies and looking after their young ones, they would not live long.

White-tail has learned to hide, to skulk from place to place without coming out in the open for his most dreaded enemy, the hunter, to see. He has learned to lurk in the shadows and to creep noiselessly through dense brush. Greatest lesson of all, white-tail has learned self-restraint. When danger threatened, white-tail's natural impulse was to run, run swiftly and far. He learned that more often than not, when he ran he got into more trouble. So instead of a mad, noisy rush through the forest, white-tail slips easily to one side, crouches low and lets danger go by.

White-tail has learned to depend on craft and cunning to escape his enemies, rather than on speed alone.

This deer has another great advantage over many other wild animals when it comes to living close to people. White-tail seems eager to be friendly. If not frightened by loud noises, quick movements or bodily injury, white-tail will soon become tame enough to walk right into the presence of people. The mother deer, if treated kindly, will bring her spotted fawns right up to the door of a mountain cabin for inspection. They are very eager to be friends.

This show of friendliness has endeared them to many people, even though a rosebush is nibbled once in awhile, or a turnip patch trampled.

BLACK-TAILED DEER

Black-tailed deer range over much of the mountainous, forested, rough, uncultivatable land in Idaho. Another name is mule deer, so called because they have enormous ears, all of ten inches long. This is a more sturdy deer than the white-tail and is generally heavier. The antler is a heavy beam that

ANTELOPE

PRONG-HORN FAMILY

(ANTILOCAPRIDAE)

The American prong-horn is known to us as the **antelope** of the Great Plains. It really is not at all like the antelope of the Old World, in fact, our antelope is in a family all by itself. The name of the family in Latin is *Antilocapridae* which in English means "antelope-goat-family." There is only one species in the whole family.

It is the only hollow-horned animal which sheds the outer shell of its horns each year. In America it is the only even-toed hoofed animal without dew-claws. It undoubtedly has the homeliest face of all the animals. Even a moose is better looking than an antelope, as far as its face is concerned. There is, however, no ugliness in an antelope's movements. It is one of the most graceful of all animals, as well as being the swiftest runner, unless you count the greyhound. The fastest race horses can run a mile in about two minutes. An antelope can easily do a mile in a minute and a half, on level ground.

The home of our antelope is the great prairies and foothills from the Cascades east to the Dakotas, Nebraska, Oklahoma and Texas. Rarely do these animals go into forests or rough mountains. They range in great herds in the states of California, Nevada, Montana, New Mexico, Oregon and Wyoming. Idaho has many thousands of these fine animals in Owyhee and Butte counties.

Antelope do not do well when confined inside fences, as in zoological parks. They seem to need lots of space for running and playing. Often a small herd of antelope will run as fast as it can for two or three miles, just for the fun of running. Barbed wire fences are a terrible hazard in antelope country. The animals do not see the wires until they are close, then they always try to run under the bottom strand. They will not try to jump, even a fence that is only four feet high. If the bottom wire is not too near the ground, antelope have no trouble crawling under. Some are sure to get cut on the sharp barbs of the wire.

Coyotes are the very worst of all antelope enemies. Bobcats and eagles are bad, too.

Antelope are increasing in numbers. In Idaho there are enough of these speedy runners to allow some of them to be taken by hunters. Hunting is closely watched by the Conservation Officers to see that not too many are killed.

HOW ANTELOPE HORNS GROW

Both male and female antelope have horns, the outer casing of which is shed each fall. The male's horns are larger than the female's and have a short prong. They grow over a cone-shaped, bony core on the frontal bone just over the eye. The core, or pedicle, is about half as long as the entire horn.

The pedicle is covered with a special pedicle skin, which is grown from the regular skin that ends at the base of the pedicle. Where the two skins meet, there is a whorl of long hair, similar to the hair on other parts of the antelope's body. The pedicle skin is covered with very fine hair.

Keratin is the substance which forms horn. It is almost exactly the same as that which makes up fingernails, hoofs, hair and the outer layer of skin. Keratin is almost one hundred per cent protein.

As soon as the outer shell of horn is shed, the pedicle skin over the bony core starts to manufacture a new one by excreting keratin. The fine hairs of the pedicle skin and some of the long hairs of the whorl become enmeshed and help to build up the lower three or four inches of horn. The upper half of the horn and the prong part are of solid keratin. Growth proceeds from a paper-thin, keratin cover of the pedicle skin to a strong, heavy, full size horn, its size depending upon the health and vigor of the antelope.

Shed horns are quickly devoured by mice, the antelope themselves, or are carried away by coyotes. They are seldom found on the range.

The horns of cattle, mountain sheep, goats and buffalo are also of keratin over a bony core, but *they are not shed*. Each year's growth leaves a visible ring at the base of the horn, which indicates quite accurately the age of the animal.

Antelope often splinter and break their horns in fighting. They never become locked together like those of deer.

HOLLOW-HORN FAMILY

(BOVIDAE)

Bovidae is Latin for "ox-family."

Buffalo are big, shaggy, black, sluggish animals that, up to about the year 1870, overran the Great Plains region of the Mississippi Valley. They were plentiful in southern Idaho. Some of the men who helped to build railroads in the West, estimated that there were sixty million of these big beasts. Others, who saw some of the herds that migrated south when winter came, estimated as high as a hundred million buffalo.

Now, all that are left are in public parks or private enclosures. The biggest number in any one herd is in Yellowstone Park. All together, there are only about four thousand in the United States.

Indians depended upon the buffalo for meat and skins. They dried the meat in the sun, powdered it between two rocks and mixed it with camas root and service berries to make pemmican for winter use. They used the skins to make tepees, clothes to wear, and robes to sleep between. But the Indians did not kill many.

Where did all of them go? Sixty million to four thousand in such a short time?

Things happen very fast in America sometimes, and the disappearance of the buffalo surely proves that rule.

But what in the world would we do with sixty million buffalo today, if we had them? Even one million, or, to be quite honest in the matter, ten thousand buffalo? Buffalo meat is good to eat, but not nearly as good as beef. Buffalo hides make very fine, warm robes, but people don't need heavy robes nowadays, as they did fifty years ago.

Buffalo have mean, dangerous dispositions and cannot be allowed to run loose. An old bull moose gets away to the most remote swamp he can find. An old bull buffalo, if turned loose, might take a notion to camp right on your doorstep.

Buffalo are very poor animals for people to try to live with. They keep well in captivity. The old buffalo you see in zoological parks seem contented to stand around day after day, lazily eating the food given them.

They are interesting to look at and study, as remnants of the once mighty herds that roamed the country where millions of people live now. Let's not worry about the passing of the buffalo. Rather, we should be glad that these few are safe behind strong fences.

THE YEAR OF THE WHITE BUFFALO

Red Plume trotted from the browning slopes of the wide valley to the Indian village beside the river, ignored the dogs that barked and children who stared, and stood before his father's wigwam, before his father, Chief Running Crane.

BUFFALO

Chief Running Crane was old. He sat on a blanket in the shade of a cottonwood tree and dreamed of the days of his youth when he killed buffalo and fished in the river for great salmon. Red Plume's lean sides heaved and there was excitement in his dark eyes as he waited for the chief to speak.

"The sun is not yet high in the heavens. You have run far and fast. You have big news, my son. Speak!"

"The valley is filled with buffalo. As far as the eye can reach, the buffalo eat grass."

"Ho! It is good. We will call the braves and make the drive. Service berries ripen on the hillsides under grey rocks. We will kill many buffalo for winter food."

"There is a *white* buffalo in the herd, my father!" Red Plume could not keep exultation from his voice.

THE YEAR OF THE WHITE BUFFALO

"A white buffalo? Are you sure, my son?" The old man could not believe his ears.

"Yes, a huge fellow. Pure white."

Chief Running Crane eyed the young man for a moment. There was new life in the aged body. He could not wholly control his voice. "A white buffalo!" he said again. Then he was silent for a long time. Red Plume stood motionless, awaiting his father's next words.

At last they came. "My father's father's father was the last of our people to see a white buffalo. It was a year of great plenty. The river was crowded with fish, the lakes covered with water birds. Berries were plentiful and the camas roots big and easily dug. Buffalo waxed fat in grass that reached to their bellies. There were deer and elk and the braves killed many of them. Our neighbors from the country of the cold wind were peaceable. There were no heavy snows. The warm winds blew and buffalo stayed in the valley all winter instead of roaming far into the land of warm winds. It was a happy hunting ground for the people, so my father's father's father said."

"And then?" Red Plume asked breathlessly.

"The braves made a big drive. They killed the white buffalo. From that time to this, snows have come to drive the herds away. Our people suffer from hunger and cold. Many die. The God of the Sun and the Warm Wind sent the beast as a reminder that, as long as it lived, the Indians should not want."

"Then we must not kill this one," said Red Plume solemnly.

A big brave stalked into the presence of the chief and stood waiting for permission to speak. He stood inches taller than Red Plume. His face was ugly with jealousy and cruelty.

"You come slowly and silently, Brother Black Cloud," said the chief. "Speak!"

The stalwart brave burst out, "The herd is in the valley. We must drive tomorow and kill for the winter. Service berries are ripe. We must delay no longer."

Chief Running Crane kept his eyes fastened on the ground. "My son, Red Plume, says there is a buffalo in the herd as white as the great swan on the lake."

"He lies! There is no white buffalo!"

Red Plume whirled on Black Cloud. Chief Running Crane

raised his hand. "No!" he said sharply. "Get my horse. I will see for myself."

Red Plume said anxiously, "You must not! You are not well. It will be too much for you."

"I will get your horse!" thundered Black Cloud. "You will see that there is no white buffalo."

Chief Running Crane, Red Plume and Black Cloud climbed a small butte well back from the river, where a good view of all the valley was to be had. As far as they could see in three directions, buffalo were scattered over the range.

Behind the men were timbered mountains. At their feet was the broad, flat stretch where they always started their buffalo drives. On the river bank was a jump-off, straight down for twenty feet. From one end of this cliff the Indians had, many years before, made a long string of rock mounds, about fifty yards apart. They were four or five feet high and wide enough for an Indian to hide behind. Carefully the Indians drove the buffalo into this trap between the rows of rock and the forest. Then, mounted on their fastest horses, with loud yells they drove the shaggy animals toward the river, shooting long arrows into the flanks of the herd to keep it running. As the herd swung toward the row of rock mounds, Indians jumped out with bows and arrows and kept the buffalo headed straight for the river.

On they went at thundering speed, their sharp hoofs cutting grass roots into dust, and the rumble of a thousand grunting, frightened beasts, crowded into a solid, rolling pack, mingled with yells from a hundred red throats and the thud of pounding feet to make a din like thunder. As the buffalo neared the jump, they tried to turn aside, but the ones behind pushed them on. Over the cliffs they went to their deaths, hundreds and hundreds of them in each drive. Then all the camp turned out to skin the big black beasts and cut the meat into strips to dry.

Old Chief Running Crane felt anew some of the excitement of the drive as he scanned the country below. He had taken part in drives ever since he was Red Plume's age, up to a few years ago when he grew too old to stand great physical exertion. Now Black Cloud led the yearly drive. Black Cloud wanted to be chief of the tribe. He would be chief when Running Crane died, were it not for Red Plume. He might be chief, anyway, if he could prove to Running Crane that Red Plume was not fit for the place.

"No white buffalo here!" the rough voice grated, as Black Cloud waved toward the feeding herd. "He has lied!"

Running Crane sat his pony and searched the landscape with eyes as keen as the eagle's. The chief's body was stiff and bent, but his eyes were as good as ever. He could see no white buffalo.

"It was right down there, near that grove of quaking aspens," Red Plume insisted, pointing with a long, lean arm.

"Are you sure, my son?"

"I am sure."

"The babbling of a child!" sneered Black Cloud. "We will make the drive tomorrow."

"And kill the white one?" Red Plume said. "Bring on deep snows, starvation and death?"

"I will see to it that the tribe has buffalo robes and pemmican before deep snows come," boasted Black Cloud, drawing himself to his full height.

"No!" exclaimed Chief Running Crane. "Red Plume does not lie. He saw a white buffalo, else he would not say so. I must see it, too, then I can go to my long sleep, where there is no hunger, no cold, no pain. Red Plume will tell his son, and his son will tell his son, and that son his son, of how Running Crane spared the herd the winter the white buffalo came. The sun will continue to shine, gentle winds will blow from the south, there will be no ice on the waters, you will not grow hungry. Running Crane has spoken!"

That night Black Cloud went about in the village, talking to the braves. "Running Crane is no longer fit to be our chief," he said. "Red Plume is not warrior enough to take part in the drive. That is why he has lied about the white buffalo. Am I not leader of the drive? Am I not the one who should be chief? Are we to starve in the deep snows because a boy tells tales that are not true?"

Next morning the braves listened again to Black Cloud. They went before the old chief.

"We want Black Cloud for chief, not Red Plume," they said. "Red Plume lies to us about a white buffalo."

"I have never known him to lie," said Chief Running Crane.

"Did you see the white one?" they asked.

"No."

"Did Black Cloud see it?"

"He says he did not."

"None of us has seen it, either. There is no white buffalo."

"Where is Red Plume?" asked the chief.

"He left the village before the sun was up."

"The sun is high. Send him to me when he returns."

Red Plume stood tall and straight before his father, when an hour had passed.

"Speak, my son!" the old man ordered.

"Again I looked upon the white buffalo. I can show him to you."

"Then get my horse. I will go. My fire is almost out. I can live but a few more days. Before I go, I want you to be chief. Show us the white buffalo and Black Cloud will be cast down by the braves."

"I cannot show it to you now. You can see it tomorrow, at sunrise."

"His talk is thin, like water!" scoffed Black Cloud. "I will not go tomorow."

He sought to dissuade the others, but when morning came a score of stalwart braves helped their old chief to mount his pony and rode with him to the top of a grass-covered ridge, six miles from the village.

The little party watched in silence as the land lighted up. They saw birds flit through low bushes, a grey wolf slink down a small coulee, a few buffalo lumber away.

As black night was washed with bright sunlight, the small band watched the sun with awe and bowed their heads in worship. It was their god and as the huge golden ball pushed up from the hill tops, they prayed to it. Their thoughts turned toward it as a god which ruled wind, rain, snow and all the wild things. They were in their temple, offering thanks to the Rising Sun.

Red Plume broke the silence. He stood a little distance apart—tall, straight, very haughty.

"There!" he said. "There is the white one."

All turned to look where he pointed. A file of buffalo wound slowly down the hill toward a small aspen grove on the creek. They were led by a snow-white, shaggy bull.

The Indians fell to the ground, all except Running Crane, who sat his horse and chanted a prayer. When it was finished they scrambled to their feet and looked again, quickly, to see if their eyes had deceived them. No! The white beast was there, his whiteness fairly glistening in the rays of the morning sun.

"Not since my father's father's father's time has a white buffalo roamed this earth," murmured old Chief Running Crane. "My son, why did we not see this great spirit of the herd before? And why has no one but you seen it until now?"

Red Plume threw wide his arms and lifted his face to the sun, to the dome of blue sky overhead.

"It is given to great chiefs and great chief's sons to be first to see the white spirit," he said. Then he dropped his chin on his chest and folded his arms.

"Come my people!" said the old chief. "We will go now to the village."

Black Cloud gathered with the rest to hear Chief Running Crane's words. He was very brief.

"This, my people, from this day forward, is your chief! Chief Red Plume!" he said and raised the boy's hand high Then the old man staggered toward the buffalo robe under the cottonwood trees.

So Red Plume became chief, not by virtue of being the biggest, strongest and bravest Indian in the tribe, but because he came from a long line of chiefs and had all the wisdom of his forefathers.

The morning before he had watched the white buffalo and some black ones wind down the hill to the aspen grove. Their bellies were full of grass and they sought the shade to chew the cud. They stayed there for an hour. When they came out, all of them were black. They had lain in a mud wallow to keep the big black flies from biting them; the black mud stayed on until it dried and the evening winds blew it away.

Red Plume wondered if Chief Running Crane had guessed.

That day Red Plume told his people, "It will be a warm winter. Food will be plentiful. The white buffalo shall live."

Then he showed again that the blood of chiefs ran in his veins and that he was fit to be their leader.

"We will make a drive!"

Black Cloud's face brightened at the words. There was reluctant admiration in it at the young chief's next words. "We will divide the herd at sunrise and drift the white one into the next valley, to be sure we do not kill it. My people shall have pemmican this winter, even though the wild swan stays and deep snows do not come. I have spoken!"

MOUNTAIN SHEEP

(OVIS CANADENSIS)

The Latin name for sheep is *Ovis*. *Canadensis,* of course, means "of Canada." The most common name for this animal is **Rocky Mountain big horn.** He has also been called the Mountaineer, and he really is an animal that loves the highest peaks and crags of our mountain country.

MOUNTAIN SHEEP

Our big horn is an ancestor of domestic sheep. Many thousands of years ago, some wild sheep were captured and tamed in far off Asia. The wild sheep came to North America by way of Bering Strait in glacial times—tame sheep came by way of Europe and England to the United States in the last three hundred years. The tame and wild sheep met on the east slopes of the Rocky Mountains. But what a difference in them now!

Big horn is tall. He stands nearly four feet high at the shoulders. His legs are long. His body is five feet long. At one end are the massive curled horns, at the other a short bobtail hardly four inches in length. The ewes are smaller and have light, slightly curved horns. The coat of Big horn is coarse, heavy hair, like that of a deer or elk. If we part that coarse hair, we find a fine wool coat underneath. If our wild sheep is a full grown buck, it weighs more than two hundred pounds. He is alert and active. His big ears are sensitive to the slightest sound and his great yellow eyes can see a long way. His nose is very sensitive, too. Here is a noble creature, at home in the highest, roughest mountains.

Now let us look at the big horn's domestic cousin:

It is short-legged, long-tailed (or it would be, if the tail were not cut off) woolly, stupid, noisy, helpless, a creature of just so much wool and mutton. A sorry-looking animal, we think, beside its wild relative.

Man has befriended and cared for the tame sheep until there are millions of them. The wild ones have been neglected and their pastures taken from them until there are very few left.

Of the enemies of big horns, probably the golden eagle is the worst. This great bird is very fond of mountain mutton and swoops down on young lambs with talons and beak. The only way the young ones can escape is to take shelter under scraggly trees, or niches in the rocks.

Coyotes, mountain lions and lynx are also vicious enemies of young sheep.

Deep snows, bitter cold and raging winds also make the big horn's life one of nip-and-tuck. Many die of starvation and disease.

The meat of mountain sheep is delicious. The horns are a highly-prized trophy of big game hunters, but hunters now must go to Canada, Alaska or Asia for one of these animals. The few that are left are protected from hunting, and the enemies of mountain sheep are being killed in an effort to give the scattered bands in Idaho a chance to increase.

Let us hope that these gallant mountaineers do multiply and prosper in coming years.

MOUNTAIN GOATS

(OREAMNOS)

The Latin name of the **mountain goat,** *Oreamnos,* means "lamb-of-the-mountains."

It is a very picturesque, pure white animal of the high western mountains. It is not as tall as a mountain sheep, but is almost as heavy. Both males and females have small, slender, slightly curved horns.

The mountain goat lives in the high mountains of Alaska and Canada, south to Montana, Idaho and Washington. The farthest south they range is the Salmon Mountains of Idaho. Just why these white mountain dwellers have not gone farther south is a mystery. Certainly it has not been for lack of mountains to travel in.

Golden eagles are the worst enemies of mountain goats. Coyotes, lynx and mountain lions also take a toll. With deep snows and severe cold to contend with in the mountains, the white goats have a tough time of it. Still they seem to do better than mountain sheep and a short hunting season is allowed in certain counties of Idaho.

Mountain goats have a coat of long shaggy hair in the winter time. This is shed when warm weather comes and the goats look as if they had just been sheared.

EARLY DAYS IN IDAHO

The emigrant train of ten wagons, some pulled by four tired horses, others by oxen just as tired, came to camp beside the river just as the sun sank behind the western hills. The wagons were pulled into a semi-circle with open side toward the clear, cold water. Horses and cattle were turned loose near a small side stream to feed on the luscious bunch grass. One of the men mounted a saddle horse and rode to a nearby butte to stand watch over the camp and stock. There might be hostile Indians in that part of the country and it was wise to be on the alert. Then too, some of the cattle and horses would start back home if they were not closely watched.

Men, women and children went about the business of making camp for the night. They gathered wood for cooking fires. They carried water in pails from the river. The women took pots and pans from the wagons, got out flour for bread and soon had it ready to bake in pans over the open fires. Men sliced steaks from the hindquarters of an elk killed a few days before. Women peeled potatoes and put hominy in a pot to boil. Others of the party busied themselves with preparing beds for the night. The women and small children usually slept in the wagons under the canvas cover. Men and boys slept on the ground under the wagons. If the night was clear the beds were often rolled out with only the moon and stars for a canopy. Some of the men looked after the harness, scraped the horses' collars clean and mended broken straps. They greased the wagons for the next day's travel and lifted heavy things about for the women. All were busy at some task or other, that is, all but John Baker.

John Baker had come back to his home in Missouri from the Civil War a very sick man. He had been wounded severely and his recovery to full health was slow. As soon as he thought he was well enough to travel he started in a covered wagon with his wife, Mary, his son, Kent and his daughter, Ann, to Oregon, where he hoped to settle in a new country. Mr. Baker had a good team, good wagon and harness, good clothes and bedding, plenty of food, three milk cows and a small amount of money.

The Baker family travelled all the way across Nebraska alone. At first, camping out at night was fun. All of them enjoyed driving over the vast, level prairie and gentle slope of hills, their eyes turning always toward the West. Soon the novelty wore off and it began to seem as if the journey would never end. Fifteen or twenty miles a day was a good average, with the cows trailing behind. Mr. Baker hunted early in the mornings, or late in the evenings, and killed antelope, deer, rabbits, geese and prairie chickens for fresh meat. They always had plenty of that, no matter what else they lacked.

At Fort Laramie, in Wyoming, they joined other West bound travellers to make up the wagon train. After they crossed Wyoming, Mr. Baker's health grew worse. He was weak and the long hard days of travel were too much for him. He hung on, however, until the night the train came to the river. That night he told the captain of the wagon train, "This Idaho country looks good to me. I am going to settle here, right here on this creek. Tomorrow we will find a place."

The others pleaded with him to go on, at least as far as the gold fields where great fortunes could be made.

"I am a farmer," said Mr. Baker. "All I want is a good place to raise crops. Goodbye to all of you, and God bless you!"

The Bakers found a fine place. There was timber nearby where good logs were plentiful for buildings. There was rich soil for crops. There were elk, deer and antelope for meat. On the river were ducks, geese and big white swans. The river teemed with fish. Grass grew high for the cows and horses to pasture in.

"Who could want a better place than this?" asked Mr. Baker.

All of them agreed that it was probably the finest place in the whole west and set about making a home with things they had brought across the plains. The first night Mr. Baker shot a deer and Kent helped him dress it and hang it in a tree.

Four years passed. One morning, just as Kent finished eating his breakfast, his father looked at him and said, "We need fresh meat. You get a deer for us today."

Kent's chest swelled a little. This was what he had been wanting to do for a long time. He had handled the long rifle, even killed geese and rabbits with it, but never had gone alone to hunt.

Ann's eyes were round with excitement. Kent's mother smiled. Kent pushed back his chair, walked to the wide fireplace, took down the powder horn from a peg in the wall, got the bullet pouch and cap box off a shelf. The long rifle rested on some deer antlers over the fireplace. He took it down, got his jacket and was ready.

Kent was very proud as he marched away with the long rifle over his shoulder. He was a man, now. Had he not been entrusted with the long rifle and sent out to kill a deer?

Up creek two miles was a likely place for deer. Kent stole to a big pine tree at the edge of a small park and searched the surroundings with keen eyes. It was an hour or more before he saw what he had come to find, a big, antlered deer, marching slowly down a path toward a saltlick near the water. It walked silently, big eyes watching for any movement in the trees or bushes. Its great mule-like ears heard every tiny sound. Its sensitive nose wrinkled as familiar odors were wafted on the warm morning breeze. To the buck there seemed to be nothing wrong, so it came slowly on toward the salty, soda-laden water of the mineral springs.

Kent stood very still. The long rifle rested butt down on the ground beside him. Although he was a tall, healthy boy of fourteen, the long rifle muzzle came even with his face. He watched every movement of the big, fat buck. It came closer. If it came to the near side of the saltlick it would not be more than twenty-five paces away. The far side was thirty-five, perhaps forty paces distant.

The buck came on. It was only sixty paces from Kent now. That was not a long shot. His father often killed deer at that distance, still, this one was coming closer all the time. Kent decided to wait a little longer. He must be very careful when he raised the long rifle or the deer would see him and bound away.

The buck must pass another tree. While its great trunk was between them, Kent raised the rifle and steadied it against the tree in front of him. His heart thumped wildly and his hands trembled as he cocked the gun and placed a fresh cap on the firing tube. The deer stepped into the open again. It looked as big as a horse to the waiting boy. Should he shoot now? Wait! something told him. The buck swung toward the near side of the lick and stopped to make sure no danger lurked there. A few steps more and it turned sideways and stopped again, as if it were suspicious.

"Now!" Kent breathed, trying to steady his trembling arms.

Overhead a pine squirrel broke out in violent chatter. It had seen the hunter and was warning the deer.

Suddenly Kent's nerves quieted. He was as cool and steady as his father would have been. He took careful aim and squeezed the trigger. The gun roared. The buck leaped into the air and bounded away. Kent's heart sank clear down to his boot tops. He had missed his first shot! What would his father say?

No, he hadn't missed! The deer stumbled and fell. It was wounded, after all. Kent took a deep breath. He must reload the rifle. He blew hard through the firing tube to be sure all tiny sparks were out. He measured powder from the horn, poured it into the muzzle, jammed down a wad of paper hard with the long, stout, hickory ram-rod. Then came the bullet which his father had cast from a pot of melted lead and finally another wad of paper rammed in hard. All ready, except for the firing cap. From a tin box he took a small, bright copper cup and fitted it over the firing tube under the hammer. He was ready to shoot again.

He could not see the buck, so he stalked carefully in a half circle to the place where he had seen it fall. It was still there —dead!

Kent stood for a long time, looking at the fine animal. "It really was too bad to kill it," he said to himself. "I like deer. I like to watch them bound along the trails through the trees. They are pretty things. I wouldn't shoot one if we didn't need fresh meat at home. Even mother, who wouldn't harm a mouse if she could help it, told father to go out and get a deer. I am glad he let me come. He said to aim for the shoulder and I did."

Kent began to feel a little proud.

"I got my buck!" he said and reached down to tug at the deer's big antlers, just out of the velvet.

He was so excited he wanted to run all the way home. He did run part of the way. Mr. Baker was busy shaping nails out of a white-hot iron bar when Kent arrived. He stood in the shop door and waited until his father looked up.

"Well, son, what luck?"

"I got a big buck, up the creek at the saltlick." Kent tried to keep pride out of his voice, to keep his chest from swelling, to act just like his father did when he had killed a deer. It was no use. Killing this first deer was a greater achievement than any other hunting would ever be.

Mr. Baker tapped lightly on the hot iron. The square nail was almost finished. A few deft blows fashioned a sturdy, square head on it. It was a good nail. Mr. Baker dropped it in a pile with some others like it on the dirt floor of the shop. He took off his apron, banked the fire in his forge and was ready to go with Kent.

"Can you and I carry your deer or shall we take the horses and stone-boat?" asked the man with a twinkle in his eye.

"We'd better take the horses for *that* buck," said Kent, and he ran to catch up the horses.

The stone-boat was a low sled built for hauling loose stones from the fields. It did not pull easily like the wagon, but the two horses were strong and the stone-boat was no trouble for them, even when the heavy deer was loaded on it. In about an hour Kent and his father were back home.

The Bakers had no railroads, no telephone, no stores nearby where they could buy things. There were no neighbors closer than ten miles. The nearest town was a long, long two days' travel with team and wagon, over a very rough road. The Bakers seldom went to town. Every spring Mr. Baker went down with furs he had caught during the winter. He always had some very fine mink, otter, lynx, beaver, wolf and bear. With the money he got for his fur he bought supplies needed at home. These were not many, as the farm, forest and river furnished nearly everything the Bakers needed.

He purchased lead for bullets, black powder, caps, iron to make such things as nails, chains and tools. He got some new-fangled steel traps, coffee, salt, sugar, medicine and cloth. Then he drove home, sleeping at night under the trees beside the road.

Most of the things the Bakers ate were raised on the farm. Corn was a very important food. Green corn on the cob, boiled and well buttered, was had in season. There was no canned corn, but Mrs. Baker cut the grains off the cobs of green corn in strips, laid them on clean white cloths in the sun and left them to dry. When well dried this corn was put away in sacks. Of course, the hard, ripe grains of corn kept all winter and through the next summer, too. Mr. Baker had a small hand mill to grind corn into meal for bread and mush. Mrs. Baker also made hominy out of corn. Still another way they liked corn was parched in the oven.

Vegetables, such as beets, turnips, onions, cabbages and potatoes were stored away in a frost-proof cellar. They had cows to furnish them with milk, butter and cheese. Chickens laid all the eggs they needed. They had pigs for lard and meat, sheep for wool. The forest furnished deer and skins, the river gave them fish, geese and ducks.

Mrs. Baker made clothes for the whole family. She spun her own yarn on a spinning wheel and knitted stockings, mittens and caps for all of them. She also taught the children their lessons.

Mr. Baker and Kent carried water in wooden buckets from the spring on the hillside. They cut wood with an axe for the fireplace. Mr. Baker made things for the house and farm in his shop on the bench or anvil. The Bakers had to take care of themselves. They were pioneers. In the four years since they crossed the plains they had built up a very fine home.

They lived well. Now Kent had killed a fat deer. It was early September and still quite warm. The meat had to be taken care of or it would spoil.

Kent and his father hung the deer from a rafter in the shop. They skinned it so carefully that when the hide was stretched out it was as clean and white inside as a sheet of paper. It would make buckskin. They cut the deer into quarters and laid them on some clean boards. The liver and heart were put in a wooden bowl. They would be eaten first.

Mother came out to help. She saw to it that every ounce of fat was saved for tallow. There was a bucket full of it.

"We shall have the saddle for roasts and steaks," she said.

"The ribs for roasts, too," said Kent. "I dearly love roasted deer ribs."

"Shall we jerk the quarters?" asked his father.

"We may as well. Kent can get us some wild geese when we need fresh meat again."

"I surely can," Kent boasted. "I'll get you another deer, too, any time you say."

Mr. Baker and his wife smiled at each other. They were proud of their older child who had grown up enough to be a deer slayer.

"Yes, I guess Kent is our hunter now," Mr. Baker said. "But let's get this deer taken care of before we kill any more."

Kent got out a keg made of oak staves and filled it half full of water. Into it he dumped a quantity of salt. This would be their brine barrel. The legs of deer meat were cut into long strips and put into the brine. After a few days they would take them out and hang them in the smoke house. Alder or mountain maple wood made the best fire for smoking and drying deer meat. When thoroughly dry and smoked the meat, now called jerked venison, or jerky, would be stored away for winter use.

In the meantime there were other tasks. A big one was making buckskin out of the deer hide. Kent started the same day he killed the deer. First he had to scrape the hair off the hide. That was not difficult when the hide was green. It was placed on a smooth log and the hair pushed off with a square-edged iron tool. In an hour both sides of the hide were clean, smooth and white.

"Run and ask Mother for the oldest, strongest butter she has," said Kent to his sister. "The older and stronger it is, the better."

She came back with a stone jar of rancid butter. Kent spread it on both sides of the hide with his hands. "Now I'll roll it up and hang it where mice and squirrels can't get at it," he said. "In three weeks I'll get it out and work on it some more."

"What a mess it will be!" Ann exclaimed, but Kent only laughed.

Next day they made candles and soap. Kent liked to make

candles of deer tallow. He selected the whitest pieces of deer fat and put them into a big, three-legged iron pot, over an outdoor fire. The fat rendered out a heavy oil which he strained through a cloth. With a dipper he filled the candle mold. The mold was of tin and held twelve cotton wicks, so twelve candles were poured at a time. When cool, the mold was held in hot water for a moment and the white, smooth, hard candles were easily slipped out. Kent and his sisters made ten dozen candles, enough to last them nearly all winter.

They also made a dressing for their shoes. This was half deer tallow and half bear tallow.

A cup full of deer tallow was saved for father to dip the bullets in when he cast them from melted lead.

All the rest of the fat was used to make soap. Bear fat, bacon grease, old lard, in fact, any and all kinds of fats were used for soap. This was not a difficult task, either. Father often helped and Mother was always around to see that the job was done right. They used the big kettle over the outdoor fire again and rendered out all the fats in it.

Lye was made by seeping water through a big box of ashes, and it was strong, too. This lye was stirred into the strained, melted fat, and the two were boiled. It took much stirring with a long, wooden paddle and smoke from the fire often smarted Kent's eyes, but he did not mind that as much as he did getting lye on his hands. It burned like fire. After a thorough boiling a very fine, soft soap resulted. For hard soap to cut into bars and lay away, the soft soap was salted out with ordinary salt. Good, solid, hard soap cooled on top of the kettle. If there had been perfume on hand, they could have made excellent toilet soap. Soap-making took a whole day, but they made enough to last for months. Sometimes, when there was plenty of fat, they made enough for a whole year.

In the next few weeks Kent and Mr. Baker were busy in the fields, gathering in the crops. Mrs. Baker and Ann were very busy taking care of them. In late September a stormy day came.

"Kent, let's make a knife today," said his father. "I bought some special knife steel when I was in town the last time. I got enough for a half dozen good knives."

"That's fine," said Kent and they went off to the shop.

Mr. Baker sat down at the grindstone with a long, narrow bar of steel in his hands. He nodded to the boy. Kent knew what his task was to be. The grindstone was a huge, round wheel of sandstone which turned in a trough of water by means of a hand crank on one side. Kent had turned the crank for his father for hours and hours at a time, mostly to sharpen axes. It seemed that axes were always getting dull. On cold mornings they brought out the iron tea kettle full of boiling water. Cold water froze too quickly and ice on the grindstone wouldn't sharpen an axe.

Mr. Baker held the steel bar on the stone to grind it into shape. This was something new to Kent.

"Why don't you heat the steel white-hot in the forge and pound it out on the anvil to the shape you want, like you do iron rings, angle irons, chains and nails?" he asked.

"Heat would take the temper out. This steel is just hard enough to sharpen and hold an edge. It came from Sweden where they really know how to temper steel. If I heat it, it will be soft like iron. Then if I try to temper it, maybe I'll get it too hard and it will be brittle. Brittle steel breaks easily. No, we must grind it down cold."

It took half a day of turning the grindstone before they had the knife blade fashioned. Although Kent's arms and shoulders ached he did not complain. He was too interested in the blade they were making.

"Now for a handle," said his father, at last. "How about stag horn?"

"Stag horn?" asked Kent.

"Yes, from your own deer."

Mr. Baker selected a tine from the antlers of the deer that Kent had killed, sawed it off at the proper length and shaped a handle from it. Filing, drilling, smoothing and fitting the handle took more time. It was almost supper time when Mr. Baker pronounced the knife finished.

"How do you like it?" he asked.

"It's a fine knife," said Kent, hefting and getting the feel of it in his hands. He tested the sharpness with the edge of his thumb. "Like a razor!" he said.

"It is for you," said Mr. Baker.

"For me!" Kent exclaimed. He could hardly believe that a

knife which had required a whole day to make could be for him.

"You can make one yourself, some day," said his father. "Ann will turn the grindstone for you."

"I must show this to mother!" Kent cried and dashed away to the house.

The next day was clear and warm. The first cold snap had passed. There would be many, many warm, clear, sparkling days before Thanksgiving. Then more snow and cold would come and winter weather begin in earnest. There was still a lot of work to do. Father intended to build a new barn. There would be logs to cut and peel. They would have to be pulled in from the woods with the team, notched and laid, one above the other, for the walls of the barn. Then smaller logs would be used for rafters and finally shakes, or clapboards, split for a roof. There was no sawmill nearby for lumber. There was no store where they could buy shingles, nails or door hinges. Mr. Baker had to make everything himself. The barn-building job would last until trapping time in January and February. Kent would help to build it.

"Before we start the barn I had better finish tanning the deer hide I put away in the shed," Kent told his father one morning.

Kent ladled a liberal supply of the soft, yellow soap they had made from wood ashes and deer fat, into a tub of warm water. He got out the greasy, smelly deer hide and rubbed it on a corrugated washboard for a half hour. It was not a nice task. Rub, rinse, clean water, more yellow soap, rub, rinse, another change of water. The skin was a heavy, sodden mass, so slippery Kent could hardly hold on to it. He soaped, rubbed and rinsed for two hours before Mr. Baker said he thought it was washed enough. "You have to break down the glue in the skin," he explained to Kent. "If it doesn't work out soft now, you will have the whole job of washing to do over."

After a final rinsing Mr. Baker helped Kent to wring the water from the slithery mass. "Now I'll help you to stretch it dry," he said.

They sat on two logs with the skin between them and pulled at it. Dinner time came. Dinner at the Baker home was at noon. Supper was in the evening. Lunch was any little snack they had between their regular meals. For dinner that day they had hot corn bread with plenty of butter, baked potatoes and fried ruffed grouse with gravy.

Before they went in to eat, Kent folded the deer skin and put it in the shop. It would not do to let it dry without stretching. After dinner he and his father resumed their task. Mrs. Baker and Ann came out to help. Pull! Stretch! Pull lengthwise! Pull sidewise! Pull! Stretch!

It seemed an endless job. Kent's hands and arms ached, but finally, as with all tedious tasks, the buckskin was dry. It was a light cream color, soft as velvet and stronger than any cloth. Kent's eyes sparkled and he forgot his throbbing muscles. He had made buckskin!

"Let's make mother a coat!" he cried. "I'll trap some mink this winter to use for trimming."

"Fine!" said Mr. Baker. "We saved the tendons out of the deer's legs for sinews. All we need to do is soak them overnight in water and split them as fine as we want. We will sew her coat with sinew. Buttons? That's easy. We'll make beauties out of deer antler."

"How about some moccasins for Ann, out of the neck part of your skin?" asked Mrs. Baker. "It is a good quarter of an inch thick and too heavy for the coat."

"Moccasins, sure! We'll make them on a stormy day when we can't work outside," Kent promised his sister.

When cold days came in December, January and February, the Baker family was cozy in their log house far up the Snake River Valley. They had plenty of food to keep them warm, good beds to sleep in, an over-abundance of good things to eat and plenty of outdoor work to keep them healthy. They were a happy, contented family. There was no time for self-pity, even if they had had good reason to feel sorry for themselves.

Deer were an important item in the lives of the pioneers. Deer furnished meat, dried meat, tallow, buckskin and even antlers for knife handles and buttons. The fur animals were important, too, for their sale brought the Bakers the only cash money they had for a long time.

RULES OF THE GAME

Why do we have laws? Why do we have laws against killing wild animals? They do not belong to anyone in particular. They live just any old place that happens to suit their fancy. Why are so many laws necessary to dictate *when* certain animals may be killed, *where* they may be hunted and *how* they may be taken?

First of all, let us go into the question of why we have laws of any kind? The answer is quite simple when we think of the things we do in life as being games.

Laws are rules of the game.

Let us think about a game that one person can play alone. Say that he is shooting at a target with a bow and arrow. He makes whatever rules suit his pleasure. He can shoot from ten paces or fifteen paces. He can shoot all his arrows as fast as he likes. There is no other person who must have a try at the target. He can laugh, shout and talk all he wishes. And these rules he makes for himself are all right. There is no one else to object.

Some friends come to visit. They want to shoot, too. They want to try their skill with bow and arrow.

Fine. Now the first fellow has some competition and rules must be established.

How far shall they stand from the target?

Shall it be rapid shooting or slow?

Who shoots first?

Shall talking and shouting be permitted?

Rules are agreed upon. "Rules of the Game" we call them and everybody abides by them. If someone is rude enough to break the rules he is unfair. He is a poor sport.

A game of baseball has many rules which the players must obey. Wouldn't ball games be funny if there were no rules? Football, tennis, cricket, basket ball and all other games are played according to rules. There could be no games if there were no rules. There would only be confused hopping about.

That's right. But what has all this to do with wildlife?

We shall see. Remember the story of Kent Baker in "Early Life in Idaho?" His family were pioneers in a land rich with deer and other wild animals. They had no neighbors for a long time. Mr. Baker or Kent could go out and kill a deer any time the family needed meat. It was an easy task. All that was necessary was to steal up creek to the saltlick and wait until a deer came down. There were plenty of deer.

When Kent ws sixteen years old, some neighbors moved in just a short way down creek from the Baker ranch. They were Mr. and Mrs. Howard, their son, Jack, and daughter, Mary. They brought a few sheep with them. The Bakers had a few sheep, too, and their cattle now made quite a herd. The Bakers were glad to have neighbors and helped them to get started on their ranch. Kent went with Jack to the saltlick and showed him how to shoot a deer. That winter Mr. Baker let Mr. Howard have half the good trapping territory to catch furs. There was plenty for both families.

Next year another family came. Mr. and Mrs. Thomas and their three children moved up creek from the Bakers and settled so near the saltlick that deer would not come to it any more. Kent, Jack and the new boy, Bob Thomas, had to find a new hunting ground quite a distance from their homes.

Now some rules for hunting deer were needed. The three families got together and agreed that Bakers would hunt on Monday and Tuesday when they wanted venison. Howards would hunt on Wednesday and Thursday, and Thomases had Friday and Saturday. No one was to hunt on Sunday. Everyone obeyed the rules. There were still plenty of deer.

As time went on more and more families came with cattle, horses and sheep. A town was founded. Many of the settlers and townspeople liked to hunt deer. They killed many of them. Everyone thought at first there would always be plenty of deer, but in four years' time the deer were beginning to get scarce. The people killed too many. They hunted whenever they could, even on Sundays.

Kent, Bob and Jack hunted for one or two days sometimes without even seeing a deer. There was none near their homes any more. Every year more people came to take up farms near there. The country became thickly settled.

More rules were needed. The old rules of the game that

had been all right for Bakers, Thomases and Howards, were not enough. If something were not done to control hunting, the deer would all be gone in two or three more years.

So a meeting was held in town. Everyone attended and there was, of course, a lot of talk about deer hunting and everything else that people talk about at such times.

This is what they said about deer hunting at the first meeting:

"We must not kill all of our deer."

"Some men hunt nearly every day."

"It is dangerous to shoot near town. Some people may be shot."

"It is a shame to kill mother deer when their little fawns are nursing. The little ones starve to death."

"Too many deer are killed when dogs are used to drive them."

"All these things are true," said John Baker, who was sort of a leader in the community. "Suppose we adopt some rules."

"Fine!" Everyone agreed.

These are the rules they adopted:

1. No rifles to be fired closer than a mile from town or anyone's ranch.
2. Hunting on week days only. No Sunday shooting.
3. No mother deer to be killed from March 1 to September 30.
4. Each family might not kill more than six deer per year.
5. No dogs should be used to chase deer.

Now, everyone was happy. There were strict rules of the game that would give deer a chance. Soon there would be plenty of deer for everyone again.

Strangely enough, deer did not increase in numbers. They became even scarcer as the years went by. Kent Baker was a grown man. Ann was a woman. Father and Mother Baker were old, grey-haired people. Another meeting was held at the town hall. It was called a sportsmen's meeting.

"People do not obey the rules of the game," said Kent Baker. "Our deer are almost wiped out. I hunted for two whole days without seeing even a fresh track of a deer."

"Something must be done," said Jack Howard. "People who break the rules in this county should be punished."

"That's right. We should make rules for punishment."

"Fine!" They all agreed. They made more rules.

1. For anyone who shot near town—five days in jail.
2. For hunting on Sunday—five days in jail.
3. For killing mother deer—ten days in jail.
4. For killing more than six deer—fifteen days in jail.
4. For hunting with dogs—fifteen days in jail.
6. The Justice-of-the-Peace, already elected by the people, would try all those who broke the rules and sentence them to jail.
7. The Sheriff, already elected, would see that the sentence was carried out.

"Now," said the sportsmen, for that is what those people called themselves, "Everyone will be fair or else be punished. There will soon be lots of deer again."

Things were a little better for one or two years.

The biggest trouble was that more people kept coming all the time. New people came from distant towns and cities, people who did not know the deer rules and did not want to. They were unfair. In our ball games such players would be ruled out. We would not play ball with them.

In the hunting game, it was not so easy to rule out the unfair ones. They sneaked into the woods and killed deer. When one of these rascals was caught and brought before the Justice, he told falsehoods. He got other rascals to lie, too. It was difficult to get the truth from such people. The Justice would not send a man to jail unless there were witnesses against him. Witnesses could not be found in many cases, so the unfair hunters went free.

Finally the people decided that there should be rigid laws passed by their State Legislature to govern deer killing. There must be paid officers to enforce these laws, too. It was not now a matter of obeying the rules of the game, like all fair people were glad to do. It was a matter of abiding by the *laws* of the State. Anyone found guilty of violating the law relating to the killing of deer was a criminal. There were special officers to bring him to justice.

Of course, there were other wild animals beside deer that required protection, and laws were passed to take care of them, too.

The laws were called game laws. Offenders were called game law violators. Officers paid to enforce the game laws were game wardens, or Conservation Officers. The *game* was the game of hunting, fishing and trapping. Nowadays we speak of deer as game. Grouse are game birds. Trout are game fish.

So, *game laws* are nothing more than the rules of the game.

Game laws are as necessary as laws against stealing, and as important.

To whom do the wild animals belong?

They belong to the states of Idaho, Washington, Oregon, Montana—whatever state they happen to be living in. And the state has the right to say what (if any), where, when and how many animals may be hunted and killed, and by whom.

Hunting, fishing and trapping were very simple matters when there were only a *few people* in the country and there was *plenty of wildlife*.

Pioneers killed all they needed.

With *more people* and *less wildlife* there had to be rules. At first they were just simple rules. Later on came stricter ones. Finally they were rigid laws, but rules, just the same.

In Idaho there is a Commission whose five members are appointed by the Governor to make the "Rules-of-the-Game" for this State. This commission meets several times each year and makes new rules as may be necessary. If there has been a drouth that has lowered the water in the streams, some new rules may be necessary for fishing. If the drouth has ruined a lot of forage, perhaps the deer hunting rules need changing. Or there may have been a terrible winter that killed many elk, so that different rules need to be made for them.

The Commission holds hearings so that any, or all, people interested may come before them and present *facts* that have to do with mammals, birds and fish. Then they make rules according to these facts.

In baseball, a committee, which is about the same as a commission, meets once a year to revise the rules of baseball, make some new ones and, perhaps, discard some old ones.

The same thing is done in college football. Once a year there are new rules. In tennis, hockey and all other games there are committees, commissions or boards to make rules. And everyone obeys these rules, too—if he wants to play.

The Fish and Game Commissioners, who have the welfare

of all wild animals at heart, make the wisest rules they know how, so that there will be hunting and fishing and also plenty of wild animals left over, to look at and admire or to hunt at some future time.

There is a purpose behind every game law, a reason for it. Laws are intended to protect wild animals from too much hunting, fishing and trapping.

A sportsman is a person who plays the game according to the rules. A sportsman never willfully breaks a rule. A sportsman never evades a rule. A sportsman never takes unfair advantage of either his fellowmen or of the wild creatures.

COUNTS, ESTIMATES AND GUESSES

Can you count the students in your school room?
That's easy, is it?
Can you count them when they are on the playground? That would be harder, but you should be able to do it with a fair degree of accuracy.

Could you count all the dogs and cats in town, be sure you had not missed any, nor counted some two or three times?

How does a sheepman count his sheep? He could not possibly count that squirming mass of woolly backs unless he put them in a corral and let them out one at a time. Cattle, too, are hard to count unless they are strung out in single file.

Every ten years our government takes a census. Counters are hired in every town and city in the whole United States to go around to everyone's home and count the people. They go into the country, to every farm and other place where people live, to get a very accurate count, so we know pretty well how many people there are.

We know how many cattle and sheep people own. No one bothers to count dogs and cats, but we can make a fair guess at how many there are in a town.

How can we count wild animals? How about mice, squirrels and gophers that live out of our sight under the ground? How about all the night prowlers? We are not very apt to see them, and how can we count what we cannot see? How about all those animals in the Far North? Those in the mountains? And all those mice, rabbits, rats and other wild creatures of the great barren deserts?

A hopeless task, you think?

You are right. It is impossible to count wild animals as you would count children in a school room, or sheep running out of a corral, one at a time. Sometimes, a little band of elk or deer may be counted. Sometimes one can tell by tracks in the snow how many animals live in certain areas. More often, however, the number of wild animals is "estimated" rather than actually counted.

Estimated?

Yes. For instance, you may count the students in one room and find there are fifty-five of them. Then you go to another school where you get just a glimpse of another room. You do not have time to count the students there, but it is a larger room than the first one and there seem to be more students in it. You estimate. You say there are about eighty-five students in that room. That is one way to do it.

Another way is to see desks, pencils, overshoes, hats and notebooks, rather than the students themselves. You estimate from all those things that there are, perhaps, eighty-five students.

Guesses? Yes, you might call them that. An estimate is a guess with some foundation in fact.

Let us estimate the numbers of some wild animals.

Take a species that are big, black and easily counted. Buffalo. We ought to be fairly sure of how many of these prairie beasts there are in the United States.

How about moose? They are big, too. But moose hide out in swamps and dense forests. We do not often see them so must take an estimate of their numbers. It could even be called a guess, and a rather wild guess, at that.

Elk? Here we must make another estimate. We can count some of them, but must guess at those we do not see.

Deer? Count part, estimate a bigger part and simply guess at most of them.

Coyotes, bears, mink and—ah! those night-time mammals! How are they counted?

To settle the whole matter, let us say we count wild animals when we can and guess the best we can on those that cannot be counted.

Let us make a list of the "estimated" wild animals in the United States:

Buffalo	4,000
Moose	14,000
Elk	200,000
Mountain sheep	13,000
Mountain goats	14,000
Antelope	165,000
Peccary	50,000

COUNTS, ESTIMATES AND GUESSES

Caribou 15
Black bear 105,000
Grizzly bear 1,000

Let's add up, just for fun. Total 566,015
A few more than half a million:
  Now for the smaller deer:
    White-tailed deer 3,578,000
    Columbian black-tailed deer 300,000
    Mule deer 1,398,000

                                                          5,274,000

Nine times as many deer as all those other animals added together. Who would have thought that?

Now let's add the number of deer to the total for all the others and we have 5,836,000. That is, less than six million of these fine animals in the whole United States.

Less than six million! Gracious! That sounds like an enormous number of big game animals. Yes, it is, but suppose we put down some more figures:

Number of big game animals 6,000,000
Number of hunters in the United States 8,000,000
Number of dogs and cats in the United
  States 60,000,000
Number of people in the United States 132,000,000
Number of sheep, cattle, horses, pigs 180,000,000
Number in squirrel family 600,000,000
Number of mice and rat family 6,000,000,000
For each big game animal there are:
    10 dogs and cats
    22 people
    30 cattle, sheep, horses, pigs, etc.
  100 squirrels, marmots, chipmunks, etc.
1,000 mice

1,162 other animals for each of those six million
       big game animals!

The eight million hunters? Not all of them are big game hunters. Only about three million have licenses to hunt deer, elk and other big game. Five million are bird and rabbit hunters.

Three million hunters for six million animals. Of the six million, ten per cent is a safe number to remove by hunting each year. That would be six hundred thousand. Six hundred thousand animals for three million hunters figures that there is one animal for every five hunters.

So, we may say with all our estimating, or guessing, if you insist, that for each deer that may be safely taken there are five hunters after it.

It is a good thing that so many hunters go into the mountains just for the trip and do not care much whether they kill a deer or not.

PART TWO

BIRDS

BIRDS

Birds are warm-blooded animals, covered with feathers, that lay eggs from which their young are hatched. Birds are flying animals, although some have strong legs and are very swift runners, like the ring-necked pheasant. Some are very fine swimmers, like the ducks. A few are divers, such as the grebe. Some make long fall and spring migrations, like Canada geese, while others are stay-at-homes, like the blue grouse. Eagles and hawks soar high to watch for prey; a woodpecker cuts a hole in solid wood to get a small worm; many birds catch live insects out of the air with great expertness; an old raven eats from a dead mammal, while a hummingbird sips of the sweetest honey.

There are birds almost everywhere on earth. Many live on the ocean. The shore is lined with them. The great forests are alive with birds. On the dry deserts one finds many species. In town, in the country, in old buildings and rock cliffs are birds. On top of the very highest mountains are ptarmigans, grouse that turn white in winter. There are owls that live in ground burrows with ground squirrels and rattlesnakes.

In the sixteen orders of Idaho birds there are about seventy-five families and probably five hundred species. Many of these are "birds of passage" and live in Idaho only while passing to and from their summering and wintering grounds. Only a relative few live in the state all year around.

Mankind has made good use of many kinds of birds. What would we do without turkeys, tame ducks and chickens? What would we do without eggs to eat?

Pheasants, grouse, quails, wild ducks and other species furnish sport, recreation and wild meat.

Insect-eating birds are helpful to the farmer. Songbirds delight us with their tunes. Prettily colored birds are a joy to behold. We love to watch and study the maneuvers of birds on the wing.

We live in a world of birds. It would be a dreary place, indeed, without them.

Of course, there are some kinds that we do not like. There are bad birds, just as there are bad mammals. We do not like owls, hawks, eagles, English sparrows, magpies and many other kinds of birds. But even they have their good points and are interesting to study.

In this book we do not have room to tell about all of the birds of Idaho. We will have to pick out just a very few species, mostly birds that are known as "game."

There are many books which describe birds in great detail as to their size, shape, color, and where they may be found. Every library has one or more bird books. All of them are worth reading and studying.

Of greatest value to you is the study of birds in the outdoors. Never be satisfied with simply knowing the name of a bird. Learn its habits. Carry a notebook and stub pencil in your pocket and write down all the things you learn about birds.

How many kinds of birds are there around your home? Do you know the names of all of them? Can you tell where they build their nests and how many eggs each one lays? Does the father bird help to build the nest and feed the young after they hatch? Do the hawks do more good than harm around your home? What are the enemies of tree nesting birds? Of those that nest on the ground?

What birds around your home are becoming more numerous every year?

Which ones are decreasing in numbers? Why?

CLASSIFICATION OF IDAHO BIRDS

CLASSES	ORDERS	FAMILIES	EXAMPLES
1. Mammals	1. Loons *Gaviiformes*	Loon	Common loon
2. Birds	2. Grebes *Colymbiformes*	Grebe	Western grebe
3. Fishes 4. Amphibians 5. Reptiles	3. Pelicans *Pelicaniformes*	Pelican	White pelican
	4. Herons, storks *Ciconiiformes*	Heron Stork Ibis	Great blue heron Bittern Glossy ibis
	5. Ducks, geese, swans *Anseriformes*	Surface feeding ducks Goose Swan	Mallard Honker Whistling swan
	6. Vultures, Hawks, Eagles *Falconiformes*	Vulture Hawk Eagle	Buzzard Redtail Golden eagle Osprey
	7. Gallinaceous birds *Galliformes*	Grouse Quail Pheasant	Ruffed Bobwhite Ringneck
	8. Cranes, Rails *Gruiformes*	Crane Coot	Little brown crane Coot
	9. Shore birds *Charadriiformes*	Plover Snipe Avocet	Killdeer Wilson's snipe Avocet
	10. Pigeons, Doves *Columbiformes*	Pigeon Dove	Domestic pigeon Mourning dove
	11. Owls *Strigiformes*	Barn owl Horned owl	Barn owl Screech owl
	12. Goat suckers *Caprimulgiformes*	Poorwill Night hawk	Poorwill Night hawk
	13. Swifts, Hummingbirds *Micropodiformes*	Swift Hummingbird	White-throated Rufous hummingbird
	14. Kingfishers *Coraciiformes*	Kingfisher	Belted kingfisher
	15. Woodpeckers *Piciformes*	Woodpecker	Flicker Redhead
	16. Perching birds *Passeriformes*	22 families Raven, swallow, magpie, crow, wren, etc.	345 species Robin Sparrow, etc.

ORDERS OF IDAHO BIRDS

ORDER 1—LOONS

(GAVIIFORMES)

Do you remember the first time you ever heard the loud laugh of a **loon**? You wondered what on earth made such a noise at night. It began low, rose high and broke off suddenly. After a minute or two the cry came again, this time like a coarse voice laughing. You may have thought that a hyena, or some other large strange animal was roaming the woods. Next day you found out it was only a bird, and after a few nights you listened for this wild "crazy as a loon" laugh. You were thrilled by it. As long as you live you will remember the cry of the loon. Have you noticed how especially noisy they are just before a storm?

The Latin name of this order is *Gaviiformes,* meaning "gorge-form," or birds that are gluttonous in their eating. Loons are large diving birds with straight, pointed bills and fully webbed feet.

There is seldom more than one pair on a lake, unless it is a very large lake. Loons are not neighborly. They drive other birds away from the vicinity of their nests. If a pair of ducks comes too close the loons have a way of diving under the water and rising from beneath the ducks to attack.

Loons feed on fish. They usually take the more sluggish ones, like squawfish and whitefish, rather than swift-moving trout. There are so few loons that the small amount of fish they eat makes little real difference to the fisherman. It is unlawful to kill loons anywhere in the United States. They are interesting birds and we wish there were more of them.

ORDER 2—GREBES

(COLYMBIFORMES)

The bird that dives more quickly than any other is **a grebe.** Often when fired upon with a shotgun it dives, just about the

GREBES

same time the trigger is pulled, and the shot passes harmlessly over the water. There are many species of grebes in Idaho waters. They are all small water birds with scalloped toes, instead of webs. Their tails are so scanty that it might be said grebes are as tailless as pika hares. Their food is frogs, fish, bugs and other small animals found in lakes and rivers.

Rednecked grebes have a black cap, white cheeks and a chestnut red neck. **Horned grebes** have a black ruff and ear tufts which give them a horned appearance. The **eared grebes** have golden plumes on the cheeks and a black, helmet-like crown. **Western grebes** are beautiful, graceful birds of the prairie sloughs. They are black and white and have long, slender necks. **Pied-bill grebes** have a stout, stubby bill with a white spot on each side. They are elusive hiders in cat-tails or other water vegetation.

Other names for grebes are: **hell-diver, water-witch, swan grebe**. The Latin name for the order is *Colymbiformes,* meaning "diver-form."

It is unlawful to kill grebes anywhere in the United States. This law was passed because the feathers of some of these beautiful birds were much sought after by milliners to trim women's hats, and too many birds were being killed.

ORDER 3—PELICANS
(PELICANIFORMES)

Pelicaniformes means "pelican-form."

The **white pelican** is the show bird of this order. **Brown pelicans** and **cormorants** are common on the Pacific Coast. Often cormorants are seen inland.

No one can mistake a white pelican. It is large. A wingspread of eight feet is not uncommon. It flies with its neck folded back against its shoulders, instead of stretched out full length like a duck or goose. When two or more pelicans fly together their wings beat in unison; as if they were soldiers marching, they keep step with their wings.

Pelicans are expert swimmers. There are three webs on each foot instead of only two, as in most other water birds. They feed on fish, frogs, snakes and any other animal life they can get. Usually it is the sluggish fish that they catch, rather than trout. They lay their eggs on barren rocky islands without making much of a nest. Young pelicans are very ugly and awkward.

We usually find a large colony of pelicans nesting together and they require peace and quiet to rear their young. Most of their nesting places have been destroyed by drainage or deliberately broken up to get rid of the big birds. Fishermen blame pelicans for the lack of fish in many streams and lakes; therefore, they persecute them wherever the birds try to nest.

A large number of these great white birds make a most spectacular sight as they wing their way back and forth from feeding grounds to nesting grounds, carrying food in their big yellow pouches to their young. Pelicans do little harm and should be protected as a most unusual form of bird life.

ORDER 4—HERON, STORK, IBIS
CICONIIFORMES)

The Latin name, *Ciconiiformes,* means "stork-form."

There are a great many species in this order. Some are very common in Idaho. Birds in this order are sometimes called "deep water waders," because of their long legs, long

BLUE HERON

necks and long bills which adapt them for feeding in rather deep water.

A good example is the **great blue heron,** often mistakenly called a fish crane. These big birds nest in rookeries, usually big cottonwood groves near the water. Both young and old are often subjected to senseless killing in the rookeries because they are supposed to be detrimental to trout streams.

The great blue heron feeds in shallow, still waters on fish, snakes, frogs, bugs and other animal life. Often they congregate at a pool left by high water, where many small fish are stranded. The fish would die anyway when the water dried up. The herons might as well eat them.

Watch a great blue heron sometime. Notice how quietly he moves about. He is a still hunter. Often he will stand on one leg for a long time, as still as a snag in the water, waiting for a whitefish, or frog to come within range of his sharp bill. Then he strikes like a flash, and he never misses, either. Try to sneak up close to a great blue heron. You will find him as wary as a coyote.

Should any bird be condemned just because Nature fashioned it to live on fish? Is it not wise that some of our fish be used by birds?

We like to see great blue herons standing along our streams. They make very pretty pictures.

ORDER 5—DUCKS, GEESE, SWANS

(ANSERIFORMES)

Anseriformes is Latin for "goose-like."

This order has a great many species, scattered wherever there is water and food for them. Birds of this order have four toes, but only two webs. They have large bills which are usually flattened. They fly with outstretched necks.

Ducks, geese and **swans** migrate north in summer and south in winter. Some are always in Idaho, both summer and winter. The summer birds go south when winter comes, others take their places. Idaho winter ducks and geese go north when summer comes and others from the south take their places. In this way some of them are always with us, and most of us are familiar with many species.

Everyone has noticed the difference in plumage between males and females of ducks. Take a drake mallard, for example: his head is green, there is pretty green on his wing,

MALLARDS

brown on his body and white on his neck and tail. The duck is more somberly colored. It is the same with other species, drakes being more brightly colored than ducks.

When summer comes, the drakes shed their bright colors and for awhile are like the females. This is called the "eclipse plumage." They cannot fly while in eclipse plumage, so they hide away in the densest part of the swamps until their wing feathers grow out.

Birds of the order of *Anseriformes* are our finest game birds. Their rivals for popularity are grouse, pheasants and quail.

Waterfowl are beautiful birds. The **wood duck, harlequin** and **ruddy duck** are among the most brilliantly colored in the whole bird class. In flight, swimming and diving waterfowl show unmatched beauty of form and motion.

The duck family is divided into five smaller families. *Anatinae* (Latin for "duck") are the river and pond ducks such as: **mallard, gadwall, widgeon, baldpate, pintail, teal, shoveller** and **wood duck.** All of these, with the exceptions of gadwall and wood duck, are very familiar to Idaho. The wood duck is the most highly colored of all ducks.

River and pond ducks feed mostly by "tipping" and reaching under the water, rather than by diving. Their food is both vegetable matter and animal life. The leaves and seeds of pondweeds, bulrushes, algae, grasses, wild millet and smartweeds are some of the principal plants eaten. These make up about four fifths of the diet. Insects, tadpoles, fish and snails make up the other fifth.

An important thing to remember about the duck family, as well as many other kinds of birds, is the great difference in size and color between males and females. Usually males are larger and prettier in color, except when in eclipse plumage.

Fuligulinae (Latin for "dusky") are the diving ducks such as: **redheads, ringnecks, canvasbacks, scaups, golden-eyes, buffleheads, harlequin** and **ruddy.**

These are the most common species in Idaho. There are some others like the **scoters, eiders, old squaw** and **Labrador duck** on the Great Lakes and sea coasts.

All ducks in this family feed by diving under water, sometimes to a depth of twenty or thirty feet to feed on plants

182 WILDLIFE OF IDAHO

MERGANSERS

CINNAMON TEAL DUCKS

growing on the bottom, or some of the thousands of kinds of small water animals.

The harlequin is a duck of swift mountain streams, often mistaken for the wood duck because of its fancy color patterns. It is sometimes called "rock duck," and mated pairs are referred to as "Lord and Lady Duck."

Merginae (Latin for "diver") are the **mergansers** or **fish-**

eating ducks. The cutting edge of the bill is saw-toothed; that gives them a firm grip on swiftly swimming trout. They also have on the end of the bill a hook which helps them hold on to their prey. Mergansers are not very desirable as a table duck, their flesh being quite strong. Young birds are sometimes eaten.

Fishermen have a very pronounced dislike for mergansers, because these birds live on trout and other fish the fishermen would like to have for themselves.

There are three kinds of mergansers: **hooded**, **common** and **red-breasted**, which make their homes in Idaho part of the year.

Anserinae (Latin for "goose") are the geese. In Idaho the **"honker"** or **grey goose**, is the most common. Honkers may be found in this state any time of the year. Many nest and rear their young along the Snake River and in northern Idaho. These probably fly south for the winter. Many geese winter in the state, too. These, no doubt, fly far north to summer. Then there are many that fly from far north to far south and stop for only a short time in Idaho.

There are several races of grey geese, some of them quite a bit smaller than the honkers, but in other ways much like them. **Brant** are a species of this family. They are small and quite dark colored. Then there is the **white-fronted goose** with a white patch over the base of its bill; the **snow goose,** a large white fellow with black wing tips; and **Ross' goose,** a small white one hardly larger than a mallard duck. The base of the bill is wrinkled, so in some regions this goose is called "scabby-nose."

Cygninae is the Latin name for **swan**. The little ones are called cygnets.

These are very large white water birds with very long necks. There are two species, **whistler** and **trumpeter.** Size is the main difference. Whistlers weigh about eighteen pounds and have a length from tip to tip of fifty-four inches. Trumpeters weigh up to thirty-six pounds and measure ten inches longer from tip to tip. A few trumpeter swans nest in Yellowstone Park, and a few more in Redrock Lake in Montana. This species is almost extinct.

Whistlers nest in the Far North; great flocks, five thousand or more, migrate south to California and Mexico for the winter.

Swans are sometimes mistaken for snow geese by hunters and killed. There is such a great difference in size between the two birds that this should never happen.

WILD GEESE ARE CALLING

"Honk! Honk! Ka-ronk!"
The sky was overcast with heavy clouds and the night as black as the inside of a tar barrel.
"Honk! Ka-honk! Kahonk!"

MELANDER AND BRANTA

There must have been a thousand geese, judging by the sound they made, and they were flying right over town. People went out on the pavement and peered upward, trying to get a glimpse of the flying wedges. No use. They might as well try to see bats in a dark cave. The arc lights threw their glare on the street. If they had been pointed up, the people might have seen the geese. They surely were close, their "Honk-honk" was just above the tree tops.

"What's happening?" someone called. The geese were

circling even lower down. There was a note of distress in their voices. They were uneasy. For hours the great birds had flown around and around, over houses and streets, and now they seemed hopelessly lost. In a great M-shaped flock, they were winging their way to their winter homes in the south. Then darkness overtook them, darkness which meant little in itself to the flock. They travelled their full speed of a mile a minute at night just as well as by day, and kept on their course just as well, too. But when the light glow of the city was reached, they lost their bearings. They circled and circled like moths around a candle flame, unable to break out of that strange light into pitch darkness again. Other flocks of geese joined them and their cries made a terrible din. One goose's loud "Honk-honk," could be heard a mile away. A dozen geese made a fine racket; a hundred sounded like a crowded street during fair week; a thousand were worse than a football game with the home team winning.

More and more people came out of their houses to stare upward into the blackness. The whole town was excited at the plight of the great birds. Their cries were not the confident, ringing lilt so pleasant to hear. They were cries of lost, confused birds, trying to find a way out of a pocket of light in a black world.

"There seems to be no hope for them but to fly all night," said a man to his neighbor. "If they become exhausted, they will fall into our yards."

"Ha, ha! Then we will have plenty of wild goose meat!" laughed his neighbor.

"A few are hitting treetops and wires now," said the first man as two big grey birds, weighing all of fifteen pounds each, toppled with broken wings in an alley.

Tom Meade was listening. All at once he got an idea, and went racing through the streets to the electric-light plant. His father was one of the engineers there. Ted burst into his presence and shouted, "Dad, can't you turn off the street lights for a few minutes 'til the geese get out of town?"

"What's that?" asked puzzled Mr. Meade.

"Can't you hear the geese? They are trapped in our lights and can't get away. Can't you hear them?"

The hum of mighty engines and dynamos in the plant

drowned out all outside noises. Mr. Meade was almost as bewildered as the geese overhead. Ted grabbed his hand and led him outside where he could hear the now exhausted birds.

"Sure I can!" said Mr. Meade when he understood the situation. "The street lights are on a separate circuit and we can turn them out."

He hurried inside, turned a big valve to slow down one of the steam engines, then pulled switches, one after another, to darken the city.

Outside, Ted listened breathlessly. Would his scheme work? Would the geese be able to line themselves out in the dark, or would they be more confused and hopeless than ever? There would still be lights from homes, and there were huge electric signs on some of the downtown buildings. Perhaps, after all, the geese would stay and circle, as the man said, until they fell dead. Ted knew that his father would not shut off the circuits that fed homes, hotels, hospitals, and business places. Anyway, it might not work to plunge the geese from light into total darkness. Ted could not detect any lessening of the frantic "Honk-honk-ka-honk!" A big goose struck some wires and fell right beyond Ted. He was beginning to feel that his idea was, after all, quite silly.

Ted's father came out to stand beside him. The telephone inside was ringing insistently, but Mr. Meade paid no attention to it. He was now as much interested in the fate of the geese as was his son.

Seconds seemed like minutes to Ted—minutes like hours. At last, there was a lessening of "Honk-ka-honk-honk!" There was distance in the noise that had been closeness only a short time before.

"They're leaving!" Ted shouted. "They're going away! They're lining out!"

"Listen to that telephone ring!" chuckled his father. "Probably the Chief of Police wants to know what is wrong with the street lights."

"Can I answer it, Dad?" asked Ted.

"Go ahead. Maybe I won't lose my job over this, and again I might," Mr. Meade muttered.

"Hello. Yes, this is the electric-light plant, Chief. Mr. Meade is busy right now—yes, he's turning the steam into one of the engines again. Yes, something happened, and he turned it off. They're coming on right now. Goodby."

Mr. Meade grinned at Ted's part of the conversation.

When the engine and dynamo were whirling again, they went outside. Far off to the south, so faint they could hardly hear it, came a "Honk! Ka-honk! Honk!" It was far away, all right, but the two listeners could tell that the old ringing notes of confidence and security was in the cry again.

"What did the Chief of Police say?" asked Mr. Meade.

"Oh, he wanted to know what was wrong with the lights! You'd think that his old jail had busted and all his prisoners had escaped, the way he talked."

"Do you suppose he heard the geese and knew they were trapped?"

"I don't know. He probably wouldn't have cared if he did. Some folks just don't understand about free things. Why, the Chief is always threatening to lock up some of us kids."

Melander and Branta escaped from the light pocket over the city with the other geese. They had summered in a small lake far north in the Caribou Mountains, and raised four beautiful goslings. These were behind somewhere in the flock formation, strong, sturdy fliers, well able to follow the leader.

Melander climbed higher in the sky. Branta, his mate, and the others followed. Free from the terrible fright of being trapped, their wings and breast muscles took on new strength. Above the clouds, they travelled under a night bright with stars. Straight as a taut bowstring they flew; just as the sun lifted over the distant mountains, they circled down into the clouds, all of them crying their wild "Honk! Ka-honk!" in anticipation of rest and food. The circle grew tighter, and as the V-shaped flock broke out of the clouds there was a large, tule-bordered lake beneath them. With a grateful "Honk!" like tired runners reaching the end of a race, they settled on the water.

Melander was a big goose. He weighed eighteen pounds and measured forty inches from the tip of his bill to the tip of his tail. Branta was scarcely smaller. They were beautiful birds. A jet-black stocking covered their long graceful necks from the pure white throat bib to the plump, grey-feathered body.

Their feet and lower legs were jet black, too. Breasts and bellies were a light cream, almost white. They held their heads high, proudly, and their eyes were keen and watchful.

All that day they rested on the lake. When evening came, the whole flock arose and scattered to near-by wheat fields to feed on grain that had been lost on the ground in harvesting. There were also many sprouted grains of wheat, and the geese ate these greedily.

Latrans, the coyote, tried to sneak up and get a fat goose for his evening meal. He saw the geese from afar and slipped like a shadow into a low gulch that bordered the field. He bellied along a fence row to a drain ditch, and padded noiselessly to a spot only a stone's throw from the feeding geese.

Latrans peeked over the bank of the ditch. His grey ears blended so well with the dead grass and wheat stubble that even the keen-eyed geese did not see him. His mouth drooled at the sight of such fat birds, so near to him and yet so far away. Latrans waited. There was always some goose with its head up, standing guard for the flock. As soon as that one put its head down, another goose's head went up. Latrans wanted to catch them with all of their heads down in the stubble picking up the wheat. Then he could make a swift run and catch one of them. But the geese were too wise to do that. Latrans thought maybe one of them would drift close to him. The birds never got close to any place where an enemy might lurk. Geese are too wily to walk into danger like that. Latrans decided to rush the flock. Perhaps one could be taken unaware. The coyote picked the time he thought best and dashed toward Melander, who happened to be closest, with mouth agape.

"Honk! Ka-hong! Ha-lunk! Honk!" the geese voiced their warnings as they made a short run on the ground and took to the air with powerful wing beats. Melander hesitated for a moment to be sure Branta was safe. Always, on the resting grounds, in the air, or on their wintering grounds, Melander looked after his mate. Let Lutra, the otter, or Corax, the raven, threaten their nest and Melander flew at them in a terrible rage. He would even fight a man if Branta was threatened. The little goslings didn't matter. Melander wouldn't fight for them, but he would give his own life, if necessary, to defend his mate.

Latrans leaped high at the grey goose. His wicked teeth snapped shut—he almost, but not quite, grasped Melander's black legs.

The honkers' flesh was tender and savoury. They fed on the seeds of water plants, wheat, barley, corn, rye, rice and any tender young plant growth. Green pasture grass and cheat were favorite foods as well as green wheat and corn. Sometimes a farmer waxed indignant because geese ate his crops, and rightly so. Of course, the geese did not know better; they fed wherever they could find good things to eat. Nature had made them that way. Hunters prized goose flesh very highly and often sought to shoot them in the farmers' fields and pastures.

Melander and Branta circled over many hunters. They knew just how high to fly above a hunter's gun so that the shot could not reach them. Once or twice, shot had pattered against their thick breast feathers and forced them higher. The honkers always circled their feeding grounds, whether it was a field of stubble or a pasture, and with keen eyes searched for enemies. A hunter had to be mighty well concealed to escape the eye of old Melander. Over their resting lake, too, the honkers always circled high out of gunshot, to settle down in the center at a safe distance. Mallards, pintails and baldpates were not so cautious and these were killed by the dozens. Once some geese were shot down when they flew too low over a long haystack where two hunters were hidden.

There was a feel of storm in the air one December day. It was not a feeling that a person could detect; it was too delicate for that, but the honkers felt it and gathered for another leg of their journey to warmer climes. They took to the air in midafternoon and formed into long flying wedges. Often one leg of the wedge would be longer than the other. Sometimes there was a double wedge, like a capital M. There were thousands and thousands of ducks in the air, too. They did not travel in military order, as if they were advancing to another front, but acted more like an army in retreat.

Melander led one great wedge into the teeth of a strong southwest wind. Pounding the air with their wings as hard as they could, they were unable to make more than thirty miles an hour against the wind. Melander went higher and still higher where the air was calm. Here they could travel their usual mile a minute. It was not good enough, so the honkers went still higher until they struck an icy blast right out of the Arctic. With this tailwind, the flock made a hundred miles an hour, nearly two miles a minute.

Long before morning came, the black-necked leader guided his flock down into peaceful balmy air. They alighted in a rice stubble field in Sacramento Valley. There was plenty of good food here, the weather was warm and the geese stayed fat. After a month or two, they made another flight south to Lake Tulare in the San Joaquin Valley, where they fed in the barley stubble fields and fresh green pastures. Here one day was much like any other day. Warm, balmy, growing days kept the honkers fat and lazy.

Melander and Branta could feel no change in the weather; in fact, there was no change. Yet there would come a time when they must go back north to nest and raise another family of goslings. Twelve hundred miles of valleys, forests, plains and mountains were between them and Kinbasket Lake, where they always made their summer home. Most of California, a wide corner of Oregon, nearly all of Idaho, and a long stretch into British Columbia must be crossed.

Why didn't they nest right there where they were and save the long trip? There was plenty of food, a nice climate and good places to nest. What was the reason for going north?

How about yourself when spring comes? Don't you get a sort of itchy feeling to be outdoors, to hike up on a hilltop and gaze out over the country, to go fishing? Isn't there an inner urge to travel? What happens when the hot days of summer come? Don't you look at the snow way up on the sides of the mountains and long to be up there? In your mind, can't you see tumbling mountain waters with trout in them? Can't you smell the pine trees? There is a special kind of warmth to a campfire at night, a keener edge to a fellow's appetite, a sounder sleep on a bed of boughs, and all this calls you to come.

So with the honkers. There was still no change in the weather where they were. Everything was just the same, day after day—except for just one thing. When Melander and Branta left Kinbasket Lake, the days were getting shorter and shorter, the nights longer. Now the reverse was true. Days lengthened, while nights shortened. So the honkers got that urge to go back to the wide, open, solitary places of the cool northland.

At first the honkers moved only a short flight each night until they got their wings trained for long trips. The muscles in the breast which pulled the great wings with such power, toughened. The wing muscles hardened, and the sense of direction quickened. Melander and the others knew the way north just as you know the way to school, only better. All wild things know directions better than people do. Your dog always knows the shortest way home, no matter where he may be. Think of homing pigeons—no amount of crooked travelling away from their homes can fool them. They always take a beeline back.

Sometimes honkers do get lost. Storms blow them off their course once in a great while. Lights of cities deceive them. But it is not often they get hopelessly off their course.

Melander and Branta got to their lake early. There were still ice and snow around the edges. Warm south winds and spring showers soon melted them away. The honkers shed their feathers. For a long while, they could not fly at all. Their nest was on a small, dry knoll, fifty yards from the water's edge.

Branta laid six eggs. Melander was always near by to fight away Corax, the raven; Crac, the crow; Mustel, the weasel; and Mephitis, the skunk. All during the nesting time, Melander stood guard. He hardly left the nesting site, even to forage for himself. Branta seldom left her nest to find food, either. At the end of four weeks, the honkers proudly marched their brood to the water, Branta leading the little ones in single file, Melander bringing up the rear. They were pretty little pale green goslings, not much bigger than an ordinary freshly hatched chicken of the barnyard. They broke file at the water's edge and plunged onto the lake among the reeds.

The old honkers watched the goslings faithfully and beat off many attacks. Yet one day a goshawk flew low and made off with one of the young ones. Melander was helpless; he could only beat his wings and hiss loudly in anger. The other goslings grew fast. The soft, fuzzy, pale green feathers gave way to hard grey ones. Long flight feathers grew out in the wings. All the time, from the very day they were hatched, they had to find their own food. Melander and Branta guided them to where the food was; they watched over them as their tiny bills picked at soft water plants, or caught water bugs. The old honkers got their feathers back, too—new, shiny, strong ones. One day in August the old ones made a short trial flight around the lake to test out their new wings. The young ones were frightened to see them take off into the air. Next day, they did not mind it so much; after a few days, the young ones tried it, too. In September all of them made practice flights over the lake. Once more, breast and wing muscles got hard and strong.

Then one blustery cold morning—the days were getting pretty short 'way up north and food was beginning to get scarce—the urge for a warmer climate was intense. "Honk! Honk! Ka-ronk!" and they were off.

The wild geese are calling. They are winging southward. Who is there so dull of mind, so wanting in love for wild things, that is not thrilled to the very core of his being by these creatures?

Twelve hundred miles south, twelve hundred miles north. "Honk! Honk! Ka-ronk!"

ORDER 6—VULTURES, HAWKS, EAGLES
(FALCONIFORMES)

Birds of this order are flesh-eaters. They have a hooked bill and four strong claws on each foot, for seizing and holding their prey. **Vultures** are mainly carrion-eaters. **Hawks** and **eagles** pursue and kill other animals for their food. **Ospreys** are often called fish-eagles.

GOSHAWK   SPARROW HAWK

The Latin name is *Falconiformes*, which is simply "falcon-like" or "hawk-like."

There is always a lot of dispute over birds of this order, as to whether they are harmful and should be destroyed, or beneficial and should be preserved. As in most disputes, there is a great deal to be said on both sides. No animal is wholly bad, whether it be a wolf, mouse, hawk, eagle, heron or fish. There is some good in almost every creature on earth. On the other hand, no animal is altogether good. Every one of them does harm to some other animal or plant.

Some kinds of hawks feed mostly on mice, ground squirrels and rabbits. Others kill many game birds and even raid the barnyard for chickens. Some hawks, such as the **rough-leg, red-tail, Swainson's** and the **marsh hawk,** are called good hawks. Marsh hawks are sometimes bad; that is, they eat too many small birds. **Chicken hawks** and the **sharp-shinned hawks** are called bad.

Bald eagles are sometimes called good. **Golden eagles** are bad, because they kill young mountain sheep and mountain goats. Vultures are good to clean up dead animals.

Before anyone starts out to kill hawks and eagles he should learn about their good points, as well as their bad ones. One man with a shotgun may do more harm killing hawks than ten men can correct by shooting ground squirrels, pocket gophers and mice.

"Be sure before you kill," is a good rule to follow.

PANDION, THE OSPREY

The ospreys came to the great canyon on Snake River in April, after the snow was gone from the rolling foothills and the long, sloping mountain ridges. The air was cool and fresh, the sun brightly warm, and the hillsides tinged with green.

OSPREYS

The first day there were ten ospreys in the canyon, putting on a glorious exhibition of aerial stunts—diving, banking, flying in spirals, climbing straight into the blue, stalling, going into dazzling pursuit and power dives, and doing everything except to fly upside down.

For hour after hour the great birds maneuvered tirelessly, showing incredible skill and amazing swiftness on their crooked, arched six-foot wingspread. As they banked and wheeled in the sun, their silvery white under parts and black upper parts flashed a dot-and-dash signal to the countryside below. "The ospreys have come back!" And all the time they screamed their love notes, "Cree-cree-cree!" They were joyful at another mating season, at being home from the far south, back to the mountain country to nest and rear their young. It was a time for courtship. They mated for life, but each springtime they renewed their pledges of love and fidelity.

The second morning there was only one pair in the canyon. Courtship was over. The rest of the ospreys had gone to other nesting places. Pandion and his mate took the old nest on top of a rock pinnacle, the aerie that had been used by their kind for centuries. The huge pile of sticks had to be made fit for nesting and rearing their young. Pandion's mate stayed in the aerie and arranged with critical care the material that he made hundreds of trips to fetch.

First he carried large sticks, some as thick as broom handles, then smaller ones, until he was bringing mere twigs. He found an old gunny sack lying beside the creek, pulled it free of entangling grass and weeds, and flew it to the aerie. A piece of rope hanging on a wire fence was a prize his mate greeted with delight. He brought dead weed stems, dried grass, paper-like leaves of mountain ash and maple, and black moss from the evergreen trees. He even carried up some old dried and whitened bones of deer. In a few days the nest was finished, habitable for another summer.

Pandion's mate laid three eggs, about the size that a large hen might lay. They were beautifully colored with purple, orange and red on a white ground, and more delicate shades —lilac, buff and peach, that deepened into full violet, red and even black. She was so proud of them she scarcely left the nest to exercise or to catch fish for herself.

Pandion was a good provider. He fished every day, most of the time in the shallow waters of Mystic Lake, four miles from the aerie, where whitefish and carp were plentiful, or in the tiny stream that threaded the canyon floor, alive with spotted trout. He never tried the river. Its waters were too turbulent for his dives.

After Pandion was settled, Haliaeetus, the enormous, white-headed eagle, whose wingspread was all of seven feet, moved into the canyon. At times his snowy crest flashed from the azure sky where he soared a mile above the earth, searching with telescopic eye for easy food. At other times he perched in a treetop, his head like a snowball in the evergreen. Baldhead's heart was as black as his head was white, where ospreys and all lesser creatures were concerned.

Life was a pleasant idyll in this picturesque country of mountains, lakes and trees, but Baldhead chafed at peace and quiet. He wanted to rob, kill and destroy. As king of birds, strongest of his kind, he was impelled to aggression against weaker species. There was plenty for all—but he wanted what someone else had.

He was bitterly jealous of the ease with which Pandion fished. It was a pretty exhibition of timing, speed and skill. The osprey came over Mystic Lake high, rode the air currents, and watched until he located a school of whitefish. He floated easily, gently, like a fluff of cottonwood down, lower, and lower, until he was only sixty feet from the water. Then, unleashing his superb strength in a burst of speed, he twisted in a long spiral down to hit the water with a tremendous splash, feet first, his long, slim, silver-lined wings folded together over his back. In a second he emerged with a two-pound whitefish firmly grasped in his pale blue talons. It weighed half as much as Pandion himself, and he needed all the energy in his lean, powerful muscles to rise from the water with it. He held the fish on an even keel, lengthwise with his line of flight, so that it looked like a part of him, like a pontoon on a flying boat.

Baldhead had watched greedily from a tall tree top in the cliffs a thousand feet above. He moved swiftly to the attack, soaring to a position four hundred feet above the osprey where he gauged its speed and dived. Pandion saw him coming and tried to hurry ahead to the aerie where he could drop the fish and defend himself, but it was too heavy to permit fast flying.

Pandion banked and spun to one side. The eagle missed, circled and climbed again. He knew now just how big the fish was and how great a handicap it was to Pandion. He dived with truer aim. Pandion banked again. The eagle snatched the fish in his own talons. There was a brief struggle, and Baldhead, twice as heavy and powerful as the osprey, flew away with the prize. Pandion alighted in a tree to rest, then went back to the lake and fished again. His mate must eat.

It was the first of a long series of piratical attacks. Baldhead hid in a different place each day, sometimes in a tall tree, sometimes in the rough cliffs that bordered the river. Often he circled out of the upper reaches of the canyon just at the right time to intercept and rob Pandion. He never tackled the osprey unless Pandion was carrying a load of fish and thus at a disadvantage. Without a handicap, in a free-for-all fight, the osprey could best old Baldhead and the eagle knew it.

One day Baldhead flew to attack with such fury that Pandion had to drop his fish in order to save his life. Baldhead swooped down, neatly picked the plummeting fish from the air and flew triumphantly to his nest with it.

Pandion sought to avoid the hijacker by going out at different times of the day. He moved his fishing grounds, from lake to creek, to another lake, and back again to the upper creek. Sometimes he was successful and won his mate's high-pitched, chirruped approval when she, instead of the hated Baldhead, got the first fresh fish of the day.

More often, however, the eagle outwitted Pandion. He watched one morning from a new lookout among the rocks, miles from the osprey's nest, as Pandion flew a circuitous route to his favorite fishing grounds. He went first to the north, as if to Mystic Lake, then swung low to the west and doubled back south to alight in some cliffs above the river. He waited there for a long time before circling slowly back. Baldhead stayed carefully out of sight and Pandion sped toward the shallow waters of the lake, glided slowly at a hundred feet until he located his prey, then went into the twisting, lightning fast spiral that ended under water with his rough, steel-like talons firmly closed on a fish.

Baldhead waited until Pandion soared high, waited until the osprey had a downhill flight to the aerie, then came up from below, forcing Pandion to go even higher. The osprey tired. His speed slackened. Baldhead darted up from beneath and grabbed the fish from the osprey's talons. Pandion had lost again.

Baldhead's success seemed to make him power mad. He grew overbold and made the grievous mistake of pursuing Pandion too close to the aerie. This time the osprey did not soar high with his catch, but flew low along the canyon floor. Baldhead came in behind him. Pandion put every ounce of strength into a final effort to outdistance his attacker. The rock pinnacle was just ahead.

"Chee-ee-eep! Chee-ee-ee-eep! Chee-ee-ee-eep!" shrilled his cry in a full-throated osprey scream.

His mate came over the rim of the nest, screaming back in a frenzy of excitement, "Chi-i-i-ick! Chi-i-ick! Chi-i-i-ick!" almost as loudly as her mate as she swooped to his rescue.

Baldhead turned in ignominious flight with Pandion's mate close upon him. The lesson did not deter the eagle in his banditry, but it made him more cautious. Next day he was back in the canyon, perched in a tall pine tree, waiting for Pandion as before, but he was very careful to stay at a respectful distance from the nesting osprey.

Three young hatched in the osprey nest, three naked, silent, awkward, ugly little things whose wide, gaping mouths spurred Pandion to redouble his fishing efforts. In a week or so they found their voices, noisy baby imitations of their parents' "Whew-whew-whew, tseep-tseep, whick-ick-ick, ku-ku-ku."

Pandion caught and carried fish to them day after day, perching on guard on the rim of the nest while he tore his catch apart with beak and talons. The mother osprey chewed the bits of carp and whitefish and placed them in the young ones' mouths.

Pandion fished for Baldhead's family, too. The old eagle robbed more persistently than ever after his own young ones hatched, although they would have gone hungry on the days that Pandion eluded him if Baldhead had not found other victims. On one occasion he robbed a pheasant's nest of all her little ones. Another time he soared to his nest with a newly born lamb in his talons and often he swooped down on some farmer's barnyard and carried away the little chicks.

On a dark, cloudy day when fishing had been poor, Pandion winged homeward with a carp much smaller than his usual catch, about a one-pounder. Baldhead saw him and dived like a streak of black lightning, his powerful, stubby wings geared to their highest speed, straight at the osprey's back.

Pandion banked sharply and the eagle shot past. This fish was so light Pandion found it no trick at all to outmaneuver the heavy Baldhead. Once, twice, three times Baldhead dived like an arrow tipped with silver, missed, flattened out, circled and climbed, only to dive again and miss. He came up from beneath, but the osprey rose more quickly. The eagle began to tire, and began, too, to lose his cool, calculating, deadly aim.

In desperation he circled high above the osprey and came down in a terrific dive, and Pandion dived, too, just a few feet ahead of the eagle. Their speed was bullet-like as they streaked downward, a hundred feet, two hundred. Pandion was spiraling, as he did when about to strike the water for a fish, but there was no water below, only trees, rocks and the unyielding earth. The bald eagle dived in a straight line.

Down and ever down they sped until Pandion seemed about to crash against the earth in his attempt to escape the robber. A scant fifty feet from the ground he leveled out and angled upward. Baldhead shot past in his straight dive, flattened out a fraction of a second too late, and with powerful wings beating the air in a frantic attempt to save himself, crashed into the top of a pine tree and hung helpless in its branches. Pandion did not falter. He headed for the aerie as fast as he could fly, the carp still locked in his talons.

The dead Eagle-of-the-Waters, Haliaeetus, his great wing bones broken, hung in the top of the pine tree beside the river. Wind, rain and sun loosened his feathers and crumbled his flesh until only a bleached skeleton dangled in the evergreen gibbet.

Free from his persecutor, Pandion found it easier to feed his family. The young ospreys feathered out and grew in size and strength. Soon came the momentous day when they left the nest on their first flight; then, as the days shortened, they made longer and longer expeditions until they were full-fledged ospreys, going into the world to catch their own fish and fight their own battles.

Pandion and his mate flew south when the first frost came, leaving the young ones to stay on at Mystic Lake and the canyon until the first thin edge of ice reached out far from shore and fishing became a lean livelihood. The young ones needed no one to show them the way south. Knowledge of that route was bred into them as much as the knowledge of fishing. When the time came, migration to Panama was as swift and certain as their twisting dives, feet first into the water after fish, osprey knowledge, like holding their catch lengthwise, to reduce air resistance.

ORDER 7—GROUSE, PHEASANTS, QUAIL

(GALLIFORMES)

These birds belong to the order of *Galliformes*, which means "hen-like" or "scratching bird."

They have four short, blunt toes for scratching; a short, horny bill for picking open the husks of grain and seeds; and

BLUE GROUSE

rather stubby, rounded wings for short flights into treetops or to help them make speedy escapes from their enemies. Their home is on the ground, although they can, and do, perch in trees. The domestic hen, guinea fowl and turkey belong to this order.

Gallinaceous birds are not migratory like waterfowl. They seldom travel far between their summer and winter homes. Sportsmen usually refer to them as upland game birds. Much of our finest sport comes from shooting upland game birds. This is a pastime that many women and girls find enjoyable.

Some species of gallinaceous birds, like ring-necked pheasants, are reared at state-owned game farms, much like domestic chickens are raised. They are then released to furnish more shooting. Ring-necked pheasant shooting is one of the most popular sports in Idaho.

Blue grouse have the Latin name of *Dendragapus obscurus* which in English is a "darksome, tree-loving bird."

Most birds, and mammals, too, move down from the high mountains to winter and go back up again to summer. Ducks and geese go from the cold north to the warm south for the wintertime.

Blue grouse are different. These big dusky-colored birds move down from the highest mountains in the spring, nest and rear their young, then go back up the mountain as far as the trees grow, to spend the winter.

Blue grouse live in the white-bark pine and fir trees during all except the most bitter storms. Then they dig under the snow. Their food is pine needles and whatever they find under dense crowned trees that the snow doesn't cover. When springtime comes, their meat is not very good to eat. It is tough and has a disagreeable pine tree taste.

Then they move down the mountainside to the aspen groves. They make a nest on the ground, along in May. Seven to ten cream-colored eggs speckled with brown are laid. The old mother grouse seems rather stupid in defending her young ones, but when we study her ways, she is not such a bad mother, after all. When danger threatens, as when a skunk appears, the little ones scatter wildly in all directions. Usually the mother flies up in a tree. When the danger is past, the mother calls her brood together again and helps them to find food. She depends on the color of her young ones blending so well with brush and shadows that the skunk will not be able to see them. At the worst, the skunk may get one, but only one —he might get two or three if they stayed huddled together. After the little ones are old enough to fly, they, too, seek safety in the trees.

Blue grouse blend so well with the shadows in a pine tree that it is difficult to see one of them on a limb. When one is disturbed, it zooms out on a swift downhill flight to the refuge of another tree, perhaps a quarter of a mile away.

During mating and nesting season the cock grouse does his best to be a dandy. He struts about with tail feathers, crest, and wings spread out widely like a turkey gobbler. On the sides of his neck are large yellow air sacs or pouches, which he can inflate and deflate at will. With pouchs fully blown up his neck is almost as large as his body. He really is a proud bird at this time. Have you ever heard the calls made by a blue grouse cock? They can be heard a mile away on a quiet morning. Sometimes it is like a "boom—boom—boom" or "whoo—whoo—whoo." Other times it may be like a small sawmill sawing knotty logs; more like groans, grunts, and grumbles than anything else. Their calls are ventriloquial; a cock booming from a tree top fifty feet above ground may sound as if it were on the ground, and a boom from the right may sound as if it were from the left. Sometimes they boom while standing on a log, and this has caused people to believe that the noise was made by beating their strutting wings on the log, instead of by deflating the air sacs. They also boom from rocks, stumps, or from the ground. They even boom while flying from one tree to another. Always the sound seems to be near by.

Food of blue grouse in summer is mostly berries, such as huckleberries, both the little red kind and the big blue ones; serviceberries, buffalo berries, bearberries, raspberries, gooseberries and strawberries. In addition, they eat grasshoppers and other insects. When fall comes, blue grouse are fat and tender. Fried to a crisp, they equal other game birds as food for people.

There is a short hunting season on grouse in Idaho, and the sportsmen kill many for the table.

Foolhen, or Franklin's grouse, smaller than the blue grouse, with white tail tips, is so foolish as to sit on a log or the low limb of a tree and let a person kill it with a stick or stones. It lives mostly in the forests on berries and mushrooms. Forest fires are tragedies for the foolhen.

There is no open season for hunters, but, of course, many mountain travellers kill them for meat, just the same.

Ruffed grouse are small fellows with a beautiful neck ruff. The males also have a big, spreading tail, of which they seem very proud.

This is a bird of the lower reaches of the streams and base of the mountains. It is often called "willow grouse," or "partridge." The ruffed grouse drums also when calling, but the sound is produced or accompanied by rapid movement

RUFFED GROUSE

of the wings which at high wing speed sounds very much like a snare drum.

In some form or another, ruffed grouse live in nearly every state in the Union, besides a large part of Canada. It is one of the best known game birds we have. There is a short open season in Idaho, during which hunters may shoot ruffed grouse. The meat is not excelled by any other bird.

Sharptail grouse, often called "prairie chickens," are a species that is very scarce in Idaho. Because they lived in an open, treeless country, sharptails lost most of their range to set-

SHARPTAIL GROUSE

tlers who plowed the land for crops or pastured it with their cattle and sheep. This was something that could not be helped. We would like to have the sharptails, but, of course, we cannot have them and people, too. It may be that small bunches will live for a long time. We cannot expect them ever to become numerous, even with the best protection we can give them.

Sage grouse, or sagehens, as they are often called, are the largest in size of all the grouse family. These are sagebrush birds, at home with the jack rabbits, antelope and coyotes. They eat the leaves of common sagebrush with relish, so much so that their flesh is always strongly flavored with sage. Even the young ones are "saged," unless killed early in August.

Because of the scarcity of these superb "cocks-of-the-

SAGE GROUSE

plains," there are not many localities where they may be legally killed any more.

Ptarmigan, sometimes called the "white-quail," is a small, rather rare, species of grouse in Idaho. Its legs, even its feet, are covered with feathers, which gives it the Latin name of *Lagopus,* or "rabbit-foot," referring to the snowshoe rabbit's foot, which is covered with hair.

Ptarmigan lives on the very highest mountains. The Forest Service reports sixteen hundred of these peculiar birds in the Nez Perce Forest. These turn white in the winter except for

and berries, and cultivated crops. These birds are sometimes very destructive to corn and other grain crops, garden crops and berry patches. On the other hand, they do much good by eating weed seeds and insect pests such as beetles, potato and squash bugs, crickets, grasshoppers, cutworms, caterpillars, moths and many others.

All birds are toothless. They do not chew their food in the mouth like mammals do.

Birds that eat grain and other coarse foods have "craws." The craw is a place in the neck for holding food before it goes on down for digestion. Into the craw go corn, wheat, or any tough-skinned insect, and with the food goes "grit." Grit may be gravel, sand, bits of broken glass, broken oyster or clam shells or any other hard, grinding material to serve as teeth for the bird. Muscles around the craw move it in a gentle rolling motion to grind the food fine, much as it would be ground in the mouth of an animal that had teeth.

Suppose snow covered the ground. Pheasants and other birds could not get grit. There would be plenty of food. Seeds of weeds, dried berries, and buds of bushes are above the snow. If someone has the welfare of the birds at heart, food will be put out for them. Piles of grain, corn and wheat will be placed in sheltered places for them to eat—but often the grit is forgotten! Might as well not put out anything for them to eat unless grit is put out, too. It is like taking the false teeth away from a person and expecting him to live on celery, tough beefsteak and hard-crusted bread. Probably as many birds die in the winter from lack of grit as from lack of food.

In some countries, pheasants are kept for ornamental birds rather than as game birds; they are kept as pets rather than as a source of food or sport.

Pheasants are extremely sensitive to distant shocks, such as explosions. It is said that in England pheasants crowed loudly and thereby warned people of the approach of German airships in the last war. In Japan they give warnings of earthquakes.

Rats, skunks, weasels, foxes, coyotes, hawks, owls, crows and—worst of all—wandering house cats, are enemies of the ringneck. Not often is a fully grown pheasant captured by these pests, but the eggs and young birds suffer great losses.

ORDERS OF IDAHO BIRDS 209

HUNGARIAN PARTRIDGES

Partridges and **quails.** *Perdicinae* is the Greek name for the "partridge-family," of which the only representative in the United States which may be hunted is the **Hungarian partridge.** This is a small, plump, greyish bird with chestnut-colored markings. It is larger than a quail, but smaller than any of the grouse. Huns, as the hunters call them, were imported to America a long while ago from Europe.

CHUKAR PARTRIDGES

Chukar partridges, and other foreign species of game birds, are being experimented with in many parts of the United States. These are released in places where there is plenty of food, water and shelter. If they like their new home they settle down, nest and raise a brood of young. Often though they are not familiar with the food plants, the enemies and other animals with which they must live, and do not survive the winters. Or they may become so homesick they die.

It is a wise plan never to kill a strange-looking bird. It may be an expensive importation, trying to adapt itself to our country.

The chukar partridge is larger than a hungarian but not quite as large as a blue grouse. It is almost the same size as a ruffed grouse. It is grey colored on its back, barred with brown and black on the upper legs and belly. There is a black stripe running backwards along the neck from the eye. It has a red bill.

There are some chukars in Idaho.

Quails are small, chicken-like, scratching birds which feed largely on insects and weed seeds. Quail are often called the "farmer's friend," because they eat so many grasshoppers, plant lice, bugs and beetles, and because they do practically no harm to the crops.

BOB WHITE QUAIL

VALLEY QUAIL

There are three kinds of quail in Idaho, all of which are fine game birds, besides being beneficial to the farmers. The **bob white** is small, not much larger than a meadow lark. It has a short round tail and no topknot. This is the bird we hear whistling, "Bob-bob-white! Bob-white! Poor-bob-white!"

MOUNTAIN QUAIL

California or **valley quail** wear a topknot of short feathers *curved forward* in a plume. They say, "Where-are-you! Chi-ca-go! You-go-'way! Qwa-quir-go!"

The **mountain quail** has a long straight topknot that sticks *straight back* over the bird's shoulders. It has a chestnut-colored throat. Its voice is a rather loud, soft, "Wook-to-wook-wook!" not often heard.

None of these three species of quail is native to Idaho; all have been brought here or have drifted here from other parts of the United States.

There are short seasons for hunting quail in some of Idaho's counties, but generally the farmers do not like to have them killed.

OPENING DAY

"Good morning, Mom," said Diana Hord, blinking sleepily as she came into the dining room. "Good morning, Dad. Why are we up so early this morning?"

"Good morning," her mother smiled.

"This is a big day, you know," said her father.

"Big day? Why?"

"Well, in the first place, it's Saturday," her father said with a twinkle in his eyes. "That means no school."

"But I usually get to sleep in on Saturday. How does it happen I'm up at five o'clock?" Diana was puzzled.

"In the second place," her father drawled, as if he had not heard her, "it's your sixteenth birthday, and we have decided to give you your present this morning."

He picked up a long package and began to untie the string.

"Oh, Dad! What is it?" Diana cried.

"You just wait a minute," he said and fumbled with the knot.

"It must be something extra special to get me up so early. Let me help you."

"No—you stand way back there. I may even have you shut your eyes for a minute."

"You'll never get those knots untied. Let me get the scissors."

"All right. My hands do seem awkward this morning."

"You aren't a little excited yourself, are you?" asked Mrs. Hord drily.

"Here are the scissors," said Diana, her eyes shining with anticipation.

"Snip, snip," went the cords on the long package. The brown paper fell away to disclose a cardboard box. Mr. Hord lifted off the top and took out a long object wrapped in many folds of paper.

"Hurry, Dad!" Diana begged. "What is it? What is it?"

"Remember last spring, when you made a perfect record skeet shooting?" asked her father.

OPENING DAY

"Yes, of course."

"Mother and I decided then to get you this for a birthday gift. Here—you finish unwrapping it." He handed the package to her.

Diana tore away the rest of the paper, and there in her hands was a brand-new, over-under shotgun. The surprise took her breath for a minute. She stared speechlessly at the beautiful little gun.

It was a twelve gauge with twenty-six-inch barrels. On one side was engraved a ringneck pheasant and on the other a ruffed grouse.

"Oh, Mom! Dad!" Diana was almost crying with joy. "Will I ever be able to knock those clay pigeons with this!"

"And that brings me to the third reason for getting you up so early this morning," said Mr. Hord.

"I can stand many more surprises like this one!" exclaimed Diana as she fondled the shotgun, unlocked the breech and peered through the shiny barrels.

"You'll have to go change your clothes," said Mr. Hord.

"What are you two planning now?" Diana looked up from the shotgun to see both of them smiling at her.

"Today is the opening of pheasant season," said her father. "We're due at the Ox-Yoke Ranch in just two hours, and it's sixty miles away."

"Now do you think we got up too early this morning?" asked her mother.

"Do you mean that I am to go pheasant hunting with you at the ranch?" Diana stared at her father with blue eyes wide.

"Yes—that is, if you want to."

"Want to! Gracious! Wait a minute and I'll be ready."

"Don't you want some breakfast?"

"I should say not! How could I possibly eat anything this morning?" and Diana was off to change into outdoor clothes.

It was just breaking dawn when Diana and her father drove into the ranch yard and switched off the headlights. They were greeted by Paul Blair, whose father owned the Ox-Yoke.

"Come in and have some coffee. We'll be ready in half an hour."

Diana took her new gun inside with her. "I've never killed

anything except a few clay pigeons," she said modestly. "I haven't even shot this gun."

"It's a peach of a gun for pheasants," said Paul. "You can't miss."

"I wish I were sure of that," said Diana.

"Just watch the dogs. They do most of the work. You can't go wrong with old David and Laurie ahead of you."

In the yard Diana looked admiringly at the Llewellyn setters. "Is it all right to pet them?" she asked.

"Sure," said Paul. "They're so excited about going hunting they won't pay much attention to you, though."

David was white, speckled with black. Laurie was white with lemon spots. Their coats were wavy, silky hair that gleamed in the pale morning light. Paul was right. They were eager to be off with the hunters, and it was only because they were polite dogs that they allowed Diana to pet them. For nearly a whole year they had waited for this day when they could pit their skill at tracking and pointing against that of the wily ringneck pheasant. They loved to retrieve the birds after they were shot, too.

"Come on," said Mr. Blair, and the four started across a hay meadow white with frost. They crawled through a wire fence into a field where white-faced cattle grazed, and on to some low, swampy ground between two gently sloping, brushy hills. They were a mile from the ranch house and had crawled through another wire fence before they loaded their guns.

"There are plenty of ringnecks close to the ranch house but we never shoot them," said Mr. Blair. "All of us like to see and hear them in the fields. There will be plenty here, I think. Paul, you take the east side and stay well up next to the trees. Mr. Hord, you can go on the west side. The sun won't bother your shooting unless you get too far ahead. Diana and I will follow the dogs straight down the middle where the cattails and swamp grass are thickest.

The sun popped over the hills by the time they were all in their places. The dogs were sent on ahead and trotted through the weeds and brush with keen noses outstretched. They could smell pheasants fifty yards away.

Diana and Mr. Blair followed about twenty paces behind, making as little noise as possible. Diana's heart was thumping wildly. She gripped her new shotgun. "Will I miss my first one? Will I miss?" she asked over and over to herself. "Dad will be so disappointed if I miss. I hardly ever miss at skeet—but this is different. This is the most exciting thing I've ever done in my life."

Mr. Blair swung over near her and motioned silently toward the dogs. All at once David stopped and stood as if frozen. His long plumed tail stuck straight out behind. His nose pointed straight ahead. One front foot lifted a little from the ground. He was pointing a pheasant!

Mr. Blair waved at Diana to go ahead. She slipped the safety catch on her gun and held it in her hands as she did for skeet while she moved slowly forward. One step, two steps, three. Nothing happened. David still stood as motionless as if he were made of iron. Where was the pheasant, anyway? Was it two feet away? Two yards? Ten yards? Which way from David was it? A dozen questions flashed through Diana's mind. She must swing with the bird when it got up. Swing her body and gun in the direction the bird flew. She must lead it a little if it went to the right or left. She hoped it would go left. It was easier to swing left than right. She must be quick. She must get her bird within thirty-five yards. Thirty would be a lot better, or twenty-five. Why didn't it fly? It must be way out in front of David. For a moment she wished that Mr. Blair or one of the others had taken this first bird. She knew all three of them were watching her. Even Laurie was watching.

Suddenly, without warning, a feathered bomb exploded in front of David. Off it went, to the right and low. "Four o'clock" was what hunters called that kind of a shot, and she had been wishing for a "nine o'clock." The bird's speed was terrific. Its wings made enough noise to shatter anyone's nerves.

Diana swung. She never knew how she managed to pivot around or how the new over-under gun flew to her shoulder. She never remembered pulling the trigger, feeling the gun kick or hearing the report, but she did see the gorgeously colored, long-tailed ringneck tumble in mid-air and drop into the cattails. There was a moment of exultation. She broke her new gun and slipped in a shell.

"That's getting them! Why, that Chink never had a chance with you and David!" exclaimed Mr. Blair.

"Fine shot!" shouted Diana's father. Paul called, "Great work!" and David plunged into the cattails and came out with the pheasant in his mouth. Admiration fairly beamed from the old dog's eyes as he laid the bird carefully at Diana's feet.

There was more shooting. David and Laurie worked fast. The ringnecks were plentiful. Laurie pointed Diana's second bird for her. It was a hen and rose more slowly than a cock, was smaller and did not cackle as it flew.

"Don't shoot that one," said Mr. Blair hastily as Diana raised her gun. She brought it down and turned to him. "There are plenty of young cocks," he explained. "We won't shoot any hens today."

Next time a cock got up it was all of forty yards away. He had taken a sneak run on David and was out of range when he flew. Again the girl held her shot.

Then she got another chance. David pointed one in a small depression. Diana was not nervous this time, and as the ringneck rocketed out she was waiting. It was a seven o'clock shot, almost straight away. The bird was hardly in the air when Diana pulled the trigger. Feathers flew in all directions as the cock tumbled.

"Give them a chance!" laughed Mr. Blair. "That one had hardly fifteen yards start. He is full of shot."

"I was afraid he'd get away," said Diana.

"Not from that new gun! I'm beginning to think that's one of those guns that never miss."

"The gun won't miss—but the person who holds it will, many, many times, don't worry!" Diana laughed back. "Today is beginner's luck."

"Bang! Bang!" went a gun as Mr. Hord shot twice for his first ringneck. The sun was bothering him.

"Bang!" Paul got a bird. And birds continued to fly for the next two hours. Everyone had three birds apiece. That was the limit. All the hunters unloaded their guns before crawling back through the first wire fence on their way to the ranch house.

"More hunters get shot crawling through fences with loaded guns than any other way," said Mr. Hord.

Diana walked beside her father toward the house. "Mother can have the pheasant dinner she likes so well," she said with a happy sigh.

"Thanks to your good shooting we can have some friends in to enjoy her dinner," said Mr. Hord approvingly.

"Thanks to David and Laurie, you mean," said Diana. "They deserve most of the credit."

"And thanks to Mr. Blair for letting us hunt on his ranch."

"There are too many thanks flying around here," said Paul. "We really do want to thank you, though, for coming."

"And to invite you to come again," said Mr. Blair. "These few days of shooting each year are the most pleasant ones of all. Bring Mrs. Hord next time. She is a fine shot—" he looked at Diana with a smile "—too!"

"Mother is getting my birthday dinner today," said Diana. She looked down at her mud-stained boots and trousers. "I'll have to hurry back to town and change to party clothes."

She got into the car beside her father and rolled down the window. "You won't miss my birthday party tonight, will you, Paul?" she asked.

"I'd just like to see anyone try to keep me away!" Paul exclaimed.

He watched the car roll out of sight along the road toward town.

"Thinking about the party, Paul?" asked his father.

"No, I was just thinking how much most girls miss by not hunting more. Didn't she have fun, though?"

ORDER 8—CRANES, MUD HENS, COOTS

(GRUIFORMES)

Gruiformes means "crane-form," although this order includes some small species like the **coot**.

Cranes are long-legged, long-necked large birds that live in marshes and feed on seeds, grain and other vegetable matter. The **whooping crane** is a very large white bird with black wing tips. It measures more than four feet from tip of bill to tip of tail and its wingspread is seven feet. These great birds are almost extinct. It would be a rare treat to see one in Idaho. The **sandhill crane** is a little smaller and grey in color. It is more numerous.

COOT AND MUD HEN

Mud hens are duck-like birds, except for rounded bill and feet without webs.

Coots have scallop-webbed toes. They are much better walkers than ducks, and often wander a long way from water looking for food. The coot's head bobs up and down when it is walking, as if it were counting every step it takes. It is a good flier, but cannot rise swiftly from the water like a duck. In the water it is a good swimmer and diver.

Nests are made of grass and weed stems and float on the water. The young leave the nest right away after hatching and can swim and dive like their parents. Thus they have a great advantage over birds that must stay in the nest a long time and be subject to attack by their enemies. Parent coots defend their nests bravely. They drive away other birds or animals that come near.

Coots are plentiful wherever there are food and water for them. It is little sport to shoot them, as they are not quick to fly. After they are shot they are poor eating. Some people have a special way of cooking coots to make them tender and savory, so it is said. After all, a lot does depend on the way wild birds are dressed and cooked. Coots eat vegetation and should not have the strong taste of fish-eating ducks, like the mergansers. And some people eat mergansers.

Ninety per cent of the coot's diet is seeds and leaves of such plants as pondweeds, algae, grasses and bur reed. (Algae is the green scum we see on stagnant pools. It is a form of plant life.) Only 10 per cent of the coot's diet is insects and other animal life.

There is a great similarity between the food eaten by ducks and that taken by coots; in fact, both kinds of birds depend on almost exactly the same food supply. Because there are so many coots in some localities, sportsmen complain that they drive the ducks away. Coots become a nuisance in other ways, too, such as damaging crops and getting into reservoirs of water upon which cities depend.

The laws allow hunters to kill a large number of coots each day during the open season, but they still increase. They have three advantages over ducks: First, they can walk better than ducks, and find part of their food on land; second, their young are able to run about, swim and dive, as soon as hatched; while ducklings must stay in the nest where their enemies have a better chance to kill them, cootlings evade their enemies; third, coots are not as desirable as ducks to the hunters, so fewer of them are shot. Is it any wonder there are so many coots?

ORDER 9—SHORE BIRDS

(CHARADRIIFORMES)

There are a great many species in the order of *Charadriiformes*, or "plover-like" birds. Some of them are: **snipes, sandpipers, plovers, killdeers, curlews, willets, yellowlegs, godwits, avocets, phalaropes** and **gulls.**

The **killdeer** is one of the best known birds of this order. All of us have seen dozens of these pretty black-and-white fellows along the shores of prairie lakes and ponds, and heard their oft-repeated cry of "Kill-de-e-er! Kill-dee-e-e! De-e-e-e!"

Curlews are often seen in moist pasture lands on their way north in the spring from the southern tip of South America. They are rather large birds with long legs and downward-curved bills.

AVOCETS

Avocets are very common in Idaho. These pretty birds frequent the shallow, alkaline lakes of the prairie, as well as shallow water along streams and larger lakes. They are rather large, long-legged birds with a long, upward-curved bill.

Altogether there are about seventy species of shore birds. They feed on vegetable and animal life found in shallow water and on swampy ground.

Birds of this order are great migrants. Many of them winter in Argentina and summer inside the Arctic Circle. The **golden plover** makes a yearly round trip that would take it almost around the world, if it travelled in one direction. It is found in Argentina from September to April. In June it is on its way to its nesting grounds in the Arctic, eight thousand miles north.

You might think that birds which winter south of the equator might go to the Antarctic for summer, but they do not. There is little migration toward the South Pole. One very good reason is the lack of food in the Antarctic. The narrow, pointed tips of both South America and Africa, next to the South Pole, are not like the vast, forested and tundra regions of North America and Asia. There is not enough room on the south end of the world for millions of migrating birds.

Migration is a law to the wild things. Wild horses, buffalo, and reindeer migrated in North America to the best areas of food. Deer and elk still migrate back and forth twice a year between summer range and winter range. Some fish migrate from fresh water to ocean, and back again. Other fish go upstream at one time of year and downstream at another. Migration is common to many kinds of wild animals. There are some things we do not understand about bird migrations. There are things we do not understand about other laws, too; for instance, why does a chick in the shell always chip a circle to the left to get out? Why not, once in awhile, chip to the right? Every unhatched bird pecks away at his shell in the same direction as clock hands move! Why?

We do not know the answer to that any more than we know why the sun rises in the east and sets in the west. It is a fixed law.

So the shore birds come and go, from 'way down in Argentina and Chile, to 'way up north in Canada. We do not know why they choose to do so, but we are glad they stop for part of the time in Idaho.

We are glad, too, that all shore birds are protected by law from shooting.

ORDER 10—PIGEONS, DOVES

(COLUMBIFORMES)

"Dove-form" birds have the Latin name of *Columbiformes*.

All of us have read about the **passenger pigeon** of the Eastern states, how there were countless millions of them, and how they all disappeared in a few years' time. Since about 1914, there has been no such bird as the passenger pigeon, except dead, stuffed ones in collections.

We have flocks of prettily colored pigeons almost everywhere. They are **domestic pigeons,** gone wild. They are descendants of the rock dove of Europe.

Homing pigeons also belong to the domestic varieties.

On the Pacific Coast are large flocks of **band-tailed pigeons** which migrate up and down the coast from California to British Columbia twice each year. These birds often do considerable damage to crops and orchards, particularly to prune orchards. There is an open season during which these pigeons may be shot.

Pigeons and doves have weak legs and feet, fit only for walking on smooth, level ground. They are not scratchers.

In Idaho the **mourning dove** is one of the most common birds, also one of the best liked. Everyone has heard its cry of "Oh-woe-woe!" as if it were really mourning some lost friend. Turtledove is another name for it. These birds are rarely found in flocks. They prefer to nest entirely alone.

In some sections mourning doves are classed as a game bird and hunted. It seems a shame to kill these harmless little birds, as of all forms of wildlife, they probably do the least harm to man and his crops.

Mourning doves feed largely on insects and weed seeds.

MOURNING DOVES

ORDER 11—OWLS

(STRIGIFORMES)

The Greek name is *Strigiformes,* or "owl-like."

> A wise old Owl sat in an oak.
> The more he saw—the less he spoke.
> The less he spoke—the more he heard.
> Why can't we be like that wise old bird?
> *Folklore*

Everyone recognizes an owl by its great round eyes and soft feathers. An owl's feathers are so soft it makes no noise in flying, as other birds do.

Nearly all owls are terrible killers of other animals, and work mostly at night. There are several species. One is the

pigmy owl that lives in hollow trees. It is not much larger than a sparrow in size, but ounce for ounce, equals the largest eagles in ferocity. This owl might be called the "weasel" of the birds.

The largest owl one is apt to see in Idaho is the **great horned owl,** known as the evil genius of the woods and barnyards. It is a great destroyer of poultry. This is the fellow who goes "Who-who-whooo!" at night.

Sometimes a **snowy owl** comes down to Idaho from the Far North. He is a big, pure white bird without horns.

Owls have very strong claws and beak. It is not safe to allow a large owl to grasp the naked fingers, as a broken bone may result.

Most owls are considered harmful to man, because they kill tame rabbits, game birds and poultry. Owls do a lot of

SHORT-EARED OWL

good in killing mice, squirrels, rats, snakes and other harmful animals. Usually, though, the bad they do is more than the good, so they are hunted down and killed in settled communities.

ORDER 12—GOAT-SUCKERS

(CAPRIMULGIFORMES)

The Latin name for this order is *Caprimulgiformes,* meaning "goat-milker-form."

Goat-suckers have flattened heads, very big mouths, small bills and very small, weak feet. They got the name of goat-sucker from their habit of flying low over goat pastures and, so people used to think, taking milk from the goats. That was a mistaken idea, as the birds feed wholly on insects and whirl through the goat pastures to catch night-flying bugs.

Because of their weak feet, birds of this order perch lengthwise on large limbs of trees, rather than crosswise on small limbs, as do birds with strong gripping feet.

There are **whippoorwills, poorwills** and **night hawks** in this order, all of whom are beneficial birds because of eating insects.

ORDER 13—SWIFTS and HUMMINGBIRDS
(MICROPODIFORMES)

These birds have the Greek name of *Micropodiformes*, meaning "small-footed-order-of-birds."

Swifts are small birds of sooty or black color, very small bill and weak, fleshy feet. These strong, fast flying birds come rightly by the name of swift. They feed on insects which they catch on the wing.

Hummingbirds are the smallest of all birds. There are dozens of species. They are the most expert of all flying animals. They can fly up, down, sideways, forward and backward. They can remain stationary in the air. No other bird can fly sideways, backward, or stay in one spot. They are great endurance flyers, too. The **ruby-throat** crosses the Gulf of Mexico in a single flight of five hundred miles while on migration.

Hummingbirds have long bills and are usually very brightly colored. They feed on nectar found in blossoms.

ORDER 14—KINGFISHERS
(CORACIIFORMES)

All of us have seen a bird somewhat larger than a robin with a great ragged blue crest, slate-blue back, white on neck and belly, sitting on a limb over a small stream, waiting

BELTED KINGFISHER

for a fish to come along so that he can pounce down upon it. We have all heard his coarse, rattling cry, too.

Kingfishers feed entirely on fish. Fishermen do not like these birds because they eat so many small fish that would later become sport for the fishermen.

This is the only species of the order of *Coraciiformes* that lives in Idaho.

ORDER 15—WOODPECKERS
(PICIDAE)

This is a very large order of **woodpeckers, flickers** and **sapsuckers.** Everyone knows one or more species as the order is widely spread over nearly all the state.

Woodpeckers have stout, chisel-like bills and strong, muscular necks for pecking holes in dead trees for worms. Their feet are very strong, too, so they can cling to upright surfaces and peck away. They have long, pointed tongues to reach into the holes for worms.

Flickers feed on ants, worms, fruit and grain. This is the big woodpecker that

FLICKER

pecks out hollow places in dead trees. Afterward, sparrow hawks, small owls, bluebirds and other birds nest in these holes.

Sapsuckers have broad, flat tongues for licking the sap from live trees after the bark has been pecked away. Their only food is sap, and they often cause great damage to orchards.

ORDER 16—PERCHING BIRDS
(PASSERIFORMES)

The Latin name is *Passeriformes,* meaning "sparrow-like." There are twenty-two families and over three hundred species, many of which are found in Idaho.

IDAHO STATE BIRD

The mountain bluebird was voted by the school children of Idaho as their favorite bird in 1929. This is one of the most beautiful bluebirds in all of the world. It is common in most sections of the state during all of the year except winter. Then it flies south to a warmer climate.

In 1931 the Idaho Legislature officially voted the mountain bluebird as the state bird.

There is another species called the Western bluebird. This one has a reddish or brownish breast. Mountain's breast is a pale blue. The females of this species are of a paler blue color with some gray feathers mixed in.

PART THREE

FISHES

FISHES

Fishes are cold-blooded vertebrates that live their whole lives in water. They breathe under water by means of gills. Their young are born from eggs outside of the parent's body, much the same as chickens are born from eggs. There are a few species of fish which bear live young, instead of laying eggs, but none of these live in Idaho.

The shape of a fish's body is like that of a torpedo. It can go through water at high speed without disturbing the quiet of the water. We say a fish's body is streamlined, so that water offers little resistance to it.

As a further aid to high speed, the outside of a fish's body is always covered with slime. It couldn't move through the water as fast, or as easily, if its body was not slick with this slime. More important still, the slime is protection against tiny plants and animals which, otherwise, would grow on the fish's body.

The main motive power of a fish comes from its side muscles which bend its tail first one way, then the other, like sculling a boat. The fins are an aid in swimming and also serve as brakes in stopping quickly.

A mammal's body is covered with hair and a bird's with feathers. Fishes have scales. They are of hard material, like fingernails, and are attached to the skin at one end. The other end overlaps the scale next to it, somewhat as shingles on a house overlap. As the fish grows larger, the scales grow too. Every year a new growth ring is added to each scale. With a magnifying glass, the rings on a scale may be counted and the age of a fish determined.

Fishes hear, see, smell and taste as other animals do. Fishes have large eyes, but no eyelids with which to cover them. Just imagine how a fish must suffer when its eyes are exposed to strong sunlight. Eyes have something to do with the color of fish. If one eye becomes blind, the fish turns dark on that side.

The fish's ears are out of sight in the head. Feeling is

SALMONOID

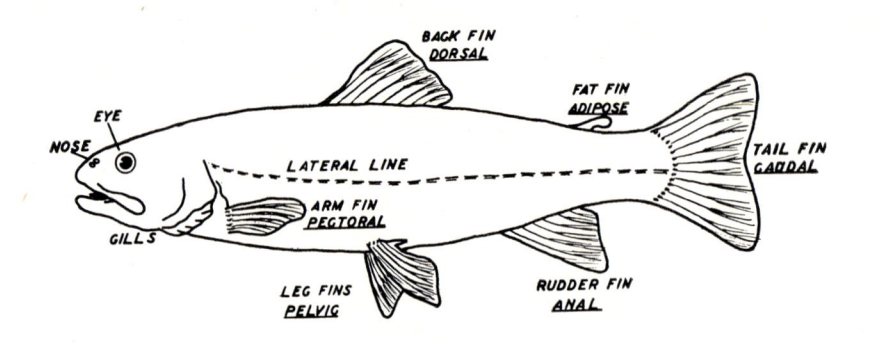

SPINY-FIN

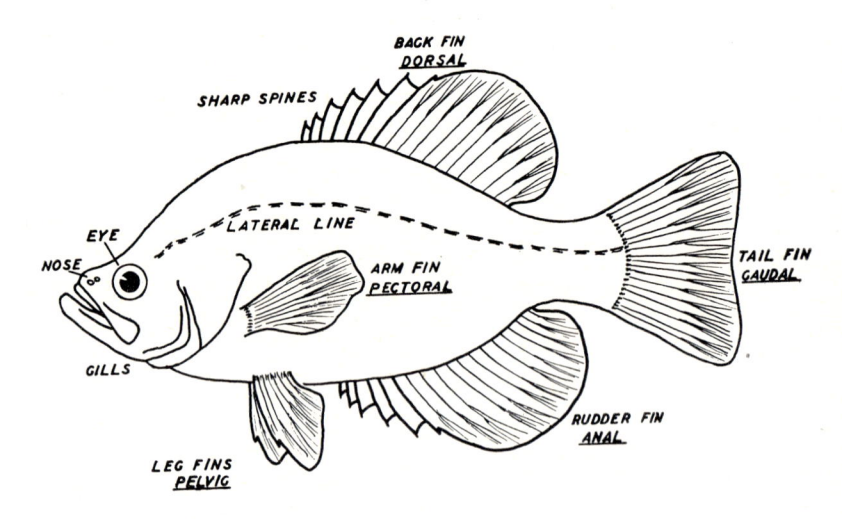

through the skin as in other animals. Taste and smell are quite keen in some fishes, rather dull in others.

The nose of a fish has two openings in front of each eye, on the outside of its body. A fish cannot breathe through its nostrils, because there are no openings to the throat, as in mammals. Water flows in and out of these openings against the sense cells.

FISHES

You have noticed the line on the sides of a trout which divides it into a top and bottom. This is called the "lateral" line. On some fishes it is crooked and higher up than on the trout. On other fishes, such as the carp, the lateral line is under the scales. Along this line is a narrow band of "sense cells," mostly the sense of hearing. Very low sounds, such as the faraway thumping of rocks together, that cannot be heard by a person, may be very distinct to a fish. These sense cells may also be a sort of thermometer, to tell very small changes in heat and cold. This lateral line may be something like a bat's nose, to warn against obstructions. It is much more highly developed in minnows than in salmon.

You have noticed the air bladder in the upper part of a fish's body. This is to keep the fish right side up. It also makes it easy for a fish to float near the surface, or to stay on the bottom, by regulation of the amount of air in the bladder.

Can a fish drown? How does it breathe?

A fish will drown about as quickly as any other animal if prevented from breathing. All animals must have oxygen for life. Mammals and birds get their oxygen out of the air through the lungs. Air is drawn into the fine cells of the lungs where it comes in contact with the blood. The blood takes up oxygen and carries it to all parts of the body.

A fish takes its oxygen from the water instead of air. There is a great deal of oxygen in water, especially water that tumbles down a mountainside or over a falls. The fish gulps a great mouthful of water rich in oxygen. Then he closes his mouth and forces the mouthful out through his gills.

Take a good look at a fish's gills. Notice the many tiny, thread-like, red-colored parts to them. All of these are full of blood and as water is forced through and across these gill parts, the blood takes up oxygen and carries it to all parts of the fish's body.

Is there ever a lack of oxygen in water?

Yes. Sometimes water stands in a pond or lake so long that it becomes oxygen-poor. We say it is stagnant and unfit for fish life. If this stagnant water were stirred up some way, like by a windstorm, there would be more oxygen in it.

Heat drives oxygen out of water, so a very warm stream may not have enough oxygen in it for fishes. Sawdust floating on top of water, sewage from cities, chemicals from factories and many other things may cause water to be unfit for fish life. Where oxygen is scarce, fishes suffocate and die.

Fish's teeth are mostly for catching and holding their prey, rather than for chewing. Most fishes swallow their food without chewing.

A pheasant, or grouse, must take grit into its craw to grind up seeds and grain before these foods can be digested. Why doesn't a fish need some such arrangement?

The difference is that plant food must be finely ground by chewing before it can be digested. Animal food, such as the fish eats, is digested without chewing. You have seen a dog swallow a big chunk of meat without chewing it very much, but you never saw a horse swallow an ear of corn without chewing it thoroughly. Fishes which eat vegetation often have gravel in their stomachs.

Fishes do not need to eat very much. One reason is that no food is needed to produce heat. The fish's temperature is always about the same as the water in which it lives. If the water is cold, the fish is cold. When the water warms up, the fish warms up, too. With a mammal or bird, it is greatly different. A deer's temperature is about a hundred and two degrees, and it must stay at a hundred and two, even if the winds blow from the north, snow falls, and the temperature gets down below zero. The deer gets enough heat from the food it eats to keep its body temperature up to a hundred and two. In cold weather, birds and mammals must eat more food, use up stored fat, or die from cold. Fish have a big advantage in not having to keep an even body temperature.

Another reason fishes do not need much food is that they are not very active most of the time. All animals become hungry when they exercise a lot. A great deal of any fish's life is spent in resting. Animals that are cold-blooded and rest most of the time can get along without much food. Often fishes become dormant and lie on the bottom of pools in much the same condition as a bear hibernating in its den. Some species, like carp, even bury themselves in mud for long periods. A few mammals, like shrews and moles, eat their full weight every day. A milk cow eats her weight in about twelve days. A fish may not require more than his weight in food every fifty days. Thus, a ground-digging mammal requires fifty times as much food as a fish of the same weight.

Nearly all fishes are cannibals. They eat the young of other species, even their own young many times. Of course, some may be worse than others in this practice, but all are bad. Sometimes fingerlings, brothers and sisters, kept in a rearing pond, will eat one another.

Fishes have many enemies. From the air they include: hawks, herons, owls, kingfishers, ospreys and even blackbirds; from land: house cats, rats, mink, otter, raccoons, bears, muskrats and skunks; and in the water are snakes, turtles, frogs and other fishes. Eels are especially bad.

Then, of course, there is the fisherman with his rod and lines to catch them from the shores. All of these put together are not as serious to fishes as the things man does to ruin the streams for fishes to live in.

We have already seen how factories, cities, and mills render streams unfit for fish life by reason of too little oxygen. Some of the other things for which mankind is to blame are: forest fires, placer mines, log drives, dams, timber cutting and irrigation.

Forest fires not only burn away the shade trees and brush, but make a lot of ashes which later wash into the streams. Fishes must have shade. They cannot live in ash-laden water.

Placer mining creates much muddy water, unfit for fishes.

Log drives not only muddy the water, but scour the stream bed of vegetation and much fish food.

Timber cutting removes necessary shade.

Irrigation ditches carry the fishes out of the streams and leave them dead on dry land. A few irrigation ditches are screened, so that fishes cannot get into them.

Dams often make obstructions in the streams that fish cannot get over in their travels to and from spawning and feeding grounds. Most dams are now provided with ladders to allow the fish to go upstream past them.

To offset all these harmful things, man has built hatcheries and rearing ponds to replenish the streams with young fish. Mankind has, in many places, destroyed the enemies of fish.

He has created huge reservoirs where fish have more water. But in spite of everything, there is less and less water suitable for fish to live in as time goes on.

CLASSIFICATION OF IDAHO FISHES

CLASSES	ORDERS	FAMILIES	EXAMPLES
1. Mammals 2. Birds 3. Fishes 4. Amphibians 5. Reptiles	1. Spiny-finned *Acanthopterygii*	*Centrarchidae* *Percidae*	Large-mouth bass Small-mouth bass Sunfish Bluegill Crappie or calico bass Perch
	2. Trout, salmon, whitefish *Isospondyli*	*Salmonidae*	Rainbow Montana black-spotted Native cutthroat Brown or Loch Leven California golden Mackinaw Eastern brook Dolly Varden Chinook Sockeye Blueback Steelhead Mountain whitefish Lake Superior whitefish Grayling
	3. Suckers and minnows *Plectospondyli*	*Catostomidae* *Cyprinidae*	Sucker Squawfish Carp Dace Red-sided minnow Stone-roller Utah chub Columbia chub
	4. Catfish *Nematognathi*	*Siluridae*	Horned pout Black bullhead Spotted channel
	5. Sculpin *Scleroparei*	*Cottidae*	Bullhead or Miller's thumb
	6. Sturgeon *Glaniostomi*	*Acipenseridae*	Sturgeon

CLASSIFICATION

Now let's take a look at the different kinds of fishes:

The *Class* of fishes is divided into orders, like the birds and mammals, also into families and species. These are in North American waters some 30 orders of fishes, 225 families and 3,300 species. In Idaho, there are 6 orders, 10 families and perhaps 40 distinct species. We shall study only those found in this state.

The first order is SPINY-FINNED fishes. Each fish of this order has, on its back, a fin with sharp bony spines in it. A person must be careful in handling bass, sunfish, perch and other spiny fishes, or he will get bad wounds in his hands.

Second order is the SALMONOID. Fishes of this order have a small, soft fin on the back, just in front of the tail, called the adipose, or fat fin. Their back fin is flexible, not spiny. Trout and Salmon belong in this order.

Third order is SUCKERS and MINNOWS. These have flexible back fins and are very bony.

Fourth order is the CATFISH. Catfish do not have scales. They have long, sharp, dangerous spines on the breast and back fins. Long feelers called "barbels" that look like whiskers are on each side of a catfish's mouth. It is these whiskers that give them their name. Catfish also have the adipose, or fat fin.

Fifth order is the SCULPINS, only one member of which is found in Idaho waters. Most of the sculpins are deep-sea fishes.

Sixth order is the STURGEONS. These are fish whose skeletons are of soft bones. They have a covering of very large, bony plates, rather than flexible scales.

ORDER OF SPINY-FINNED FISHES

Family of Bass and Sunfish
Family of Perch

(ACANTHOPTERYGII)

Acanthopterygii is the Greek name for "spines-in-the-fins," and refers to the hard, horny, needle sharp spikes that make a fish of this order so disagreeable to handle. There are from five to thirty-seven of these spines in the one, or two, back fins, which may be smoothed down in a groove on the fish's back. If rubbed the wrong way, they stand erect. There are also sharp spines in the rudder fin.

Spiny-fins are usually prettily colored, often with bright greens and yellows. They are sporty fishes and very good to eat.

BASS AND SUNFISH FAMILY

These fishes dig out a big nest in the springtime at the side of a pool in which the eggs are laid. Nature has provided a sticky substance on fish eggs so they will readily adhere to the gravel and other material in the nest and not be washed away. They must be in continual motion to keep them from sticking together and thus allowing a fungus growth to start. The eggs must also be kept clean and free from silt and dirt, for the eggs, like the fish itself must have oxygen.

The mold on sour milk, or cheese, which has stood too long in a warm place is a fungus. We have all seen green, moldy bread; this is a different kind of fungus. And fish eggs, and even the fish themselves, are troubled with still other kinds. Eggs that stick together and mold are spoiled.

The father fish stands guard on the nest until the young fish hatch. He fights angrily any other fish that comes near. Besides, it is his duty to stir the water on the nest with his fins to keep the eggs from sticking together.

All the time the father bass, or the father sunfish, guards the nest he goes without eating and, of course becomes very hungry. His disposition gets worse and worse. He fights any and all things that come along.

Do you think a bass would be easy to catch on a hook while he is at the nest?

Yes, he is very easily caught while hungry and mad.

What happens to the eggs if the father fish is caught?

All of them mold or get covered with mud and become spoiled. The whole clutch is ruined.

Then it seems poor practice to catch bass or sunfish while they are nesting?

Yes. Bass or sunfish should never be caught at this time of year.

There are two kinds of bass: *Large* mouth and *Small* mouth. Their nesting and other habits are quite alike and there is even a great similarity in their appearance. Because small mouth bass have just been planted within the past few

SMALLMOUTH BASS

LARGEMOUTH BASS

lures, with a savage fierceness; when hooked they put up a good fight.

A string of bass fried to just the right crispness, oh, my, but they are good! And what a sport it was to catch them!

Sunfish. There are two kinds in Idaho: **common** and **blue**. They are also called "pumpkin seeds," because they are shaped something like a pumpkin seed; larger of course and colored differently.

The common sunfish have ten or twelve spines in their back fins and three in the rudder. They live in lakes, ponds, and streams which do not flow swiftly. The little ones are transparent, so are invisible in water.

This is a very fine fish. It is not difficult to catch and has a fine flavor when cooked. It has a tender mouth which is easily torn if the hook is jerked too hard.

Blue sunfish or bluegills have a tiny dark blue, almost black spot on the gill cover. On the common sunfish this spot is deep orange, or red colored. Blues have the same number of spines as the common sunfish.

Crappies are often called sunfish or pumpkin seeds. Crappies are usually larger than sunfish; sometimes one may be caught that weighs two pounds and measures thirteen inches long and five inches deep from back to belly. The average size, though, is almost like a man's hand. Crappies are a sort of transparent silver color with black spots. Often when you first look at a crappie, you think you might hold it up to the light and see right through it. Crappies have five or six spines in their back fin.

Calico bass. This is commonly also referred to as a crappie and is similar in appearance. The calico bass has six or seven spines in the back fin. This is the species most common in Idaho.

Sunfish and crappies guard their nests the same as bass do.

Perch belong to a different family from the bass and sunfish. The **yellow perch**, found in lakes, ponds and warmer waters of Idaho, is not as wide and flat as sunfish and crappies. The head is rather long instead of being snubbed off like that of a sunfish. Their color is orange or yellow with prominent black vertical tiger stripes.

They have two prominent back fins, twelve to fifteen spines

BLUEGILL SUNFISH

CRAPPIE, OR CALICO BASS

JUST FISHIN'

"Let's go fishin'!"
"Good idea. Where shall we go?"
"To the lake. The sunfish and perch are bitin' good."
"O. K. Got any bait?"
"We'll dig a few angleworms."
"Got a pole?"
"No. We'll cut a willow."
"I've got two hooks and a long line that Dad gave me."
"That's all we need. Come on. My mother will like some nice fat sunfish."

Just like that, you arrange for a fishing trip and walk down to the small lake, cut a long willow pole apiece, tie on the line and hook, bait with a small piece of angleworm and throw in.

A nibble! A bite! You have one hooked and throw it out on the shore. A nice fat perch flops in the bright sunshine. While you are getting it off the hook, your partner pulls out a sunfish, even more prettily colored than your perch. You stop only a moment to admire their beauty before throwing your line back into the water for another bite and another fish.

Soon you have a dozen of them lying on the bank. You cut a slim, forked willow and string your fish on one side of it, lay them in the edge of the water and put a rock on the end of the willow to keep it from floating away. Now you can skim rocks for awhile, climb a tree or two where there are bird nests to look at, go swimming and make some whistles out of elder-brush stems. Back where you left your fish is an old Bull snake trying to get a fish off your willow fork. You chase him away, get your fish and go home with meat for the family dinner. You feel mighty proud, too, that you have brought home food for the others; just as grown up as if it were a couple of cotton-tail rabbits, or some China pheasants. And you have had fun, too!

Next time you have a half holiday, you go out for catfish. These lazy fellows stay on the bottom, and you must put a

JUST FISHIN'

sinker on your line to get the hook low enough for them. Then you need a bobber, too, a float to keep the hook, baited with an angleworm, from sinking into the mud. With a little experimenting you can find just where to slide the bobber on the line so that the hook will be only an inch or two from the mud. You toss in and wait. Catfish are not very energetic in feeding. If one drifts along and finds your bait, he is apt just to touch it with his mouth, just enough to make your bobber tremble a bit. Next he nibbles at it, the bobber goes up and down, you have a bite! All at once the bobber goes out of sight under the water—your fish has swallowed the bait and is running. You give a pull on the pole and out comes your catfish, a nice one, ten inches long. Be careful how you take hold of it—remember those sharp, long daggers in the back and arm fins.

Catfish are biting slowly this day so you make a set. You cut a willow fork about three feet long to stick down in the mud for the pole to rest on. Then you shove the end of your pole in the ground, or else lay some rocks on it, toss your line out and go away. If a catfish bites, he will hook himself and be there when you return. Don't stay away any longer than an hour, though, because the Idaho fishing laws state that a line must not be left unattended for more than one hour.

Catfishing is slow work, not at all exciting, but you catch six nice ones, and, my, how good they are to eat! You have had fun, too. You had lots of time to think about things. That new shotgun, for instance, or maybe a pony to ride to school. Your dog was along and you had a wonderful time helping him to dig a rabbit out of a hole. The rabbit got away, Dandy was a little slow in making a start when bunny jumped out, but Dandy chased it a mile or more. Fishing is good for the brain. Everything seems to sort of smooth out for a fellow when he goes fishing.

Next time your fishing is in the mountains for trout. You have a light, flexible, jointed rod, very pretty with colored silk thread wrappings, nickel-plated bindings, a cork handle, and, of all things, a reel on which your line is wound. You can have a long line or a short one simply by "reeling out" or "reeling in." The line runs through guides to the tip. On the end of your line is a "leader" made of light-colored material that is invisible when wet. The hook has a few colored bits of feathers tied on it and we call it a fly. After all this equipment is rigged, we sling a creel on our shoulders and are ready. The creel is a basket to hold our fish.

Now, shall we fish downstream or upstream? We are after trout. Trout always lie with their heads upstream. Food floats down to them, seldom up. All right, we go upstream casting our fly on the water, pulling it back and casting again, time after time. The idea is, some trout will see the fly alight, think it is a real fly and strike it. Trout do not bite, they strike, in fishing language. Well, we whip the creek for a long while, maybe a half hour or more, before we come to a deep, shady pool under a rock cliff. Ah! Your interest quickens. That is a likely place. You pause to change your fly. The one you have on is grey and red. You put on a brown, white and red one, slip up cautiously and whip it to the upper end of the pool. White water takes it away for a moment, then you see it just under the surface, and you see something else, too. A long, silvery shape darts at it, clear out of the water flashes a foot-long rainbow as it strikes savagely at your fly. Give your rod a light jerk. Don't strike too quickly, or you will take the fly right out of his mouth. Give him time to turn so the hook will pull toward the fish. But you must set the hook in his jaw before he finds he is fooled and spits out the feathers he thought were food.

Now you have him hooked. You find him running, dodging, twisting, like a quarterback trying to run back a punt. You have your hands full, reeling in when you can, letting him have more line when he runs away. You cannot pull him out on the shore as you did a catfish. If you try, you will break your rod or line, so you must tire this fellow out and gradually pull him to shore. It is as exciting as that time a horse bucked with you, or when you took your first airplane ride. This is really fun— and how good that trout will be when fried!

Eagerly you go on to the next pool, catch two out of that one, and three from the third. When evening comes you have eight fine fish, all of them a foot long or more.

When you get home and lay your fish in the kitchen sink, don't you sort of swell up inside and feel as if you've brought home just about that much gold? Fishing is really great. Great recreation—great for getting some good food, too.

One day we go to the city and see a crowd of men and women fishing at one end of the wading pool in the park.

"They must be crazy."

"There's no fish in there. Why, it isn't over two feet deep!"

"Let's go over and watch them."

"O. K."

We go over to this strange gathering of men and women. They do not look crazy, still they are rather queer looking, at that. They seem awfully serious and hardly one of them ever smiles.

In the wading pool are anchored some auto tire casings, and as we watch we find that the people are trying to place their flies inside one or the other of those casings. They have long, very flexible rods. We watch one man cast. He is a great big man with a long face, and long arms.

He whips his line with as much seriousness as our organist sits down to play at church. We watch breathlessly. The thin line cuts the air in beautiful loops, and the fly falls exactly in the center of the target.

"Ah!"

"Oh!"

"Perfect!"

"Grand!"

The watchers look with awe and wonder at the caster.

Again he makes a cast, taking plenty of time to get ready and be sure every part of his outfit is in perfect order. He moves his feet, like a golfer taking stance for a difficult shot. With dignity befitting a band leader he moves his rod again, back and forth, the tip making larger and larger curves, the line moving in larger and more precise loops until with one final, grand swing of his arm, the fly is thrown out to land, as before, exactly in the center of the the old tire.

Again the others show how they admire and worship such skill, with many "Ah's, Oh's, Grand's, Magnificent's, Glorious's."

"You are the champion," they tell him. "You have not missed today." And they give him a blue ribbon.

"Say, mister," we ask a rather friendly-looking fellow at one end of the crowd. "Would you mind telling what kind of fishing this is?"

"I don't mind at all. Let's go over and sit down on a bench. I'm tired of standing."

"You see—" this man has a twinkle in his eye that makes a fellow like him, right off "—fly-casting is an art, just like painting, music, sculpture, or—" he sort of hesitates a bit "—or fishing. These men and women do wonderful things with a fly rod, silk line and fly. They have a good time without ever thinking of catching a fish. Why, they can practice fly-casting in their back yards. See that red marker over there? Well, a man stood today, where you saw the caster stand, and put a fly down at that marker."

"Gee whiz! That's as far as a fellow can throw a baseball—almost." It begins to look as if maybe there is something to this funny-looking business after all.

"And you saw a man drop his fly into those hoops. It takes a lot of practice to be able to do that. Practice and skill."

That was quite a stunt. Lots of times when we try to put our fly in a certain spot, it goes somewhere else and often gets tangled in brush or trees.

"I'll bet you could do as well, mister."

"Oh, no! I couldn't hit even the closest hoop."

"I'll bet you, though, that you can catch fish."

"Well—" the man is rather bashful "—I did catch a cutthroat once that weighed eleven pounds," he admits, and we can see his chest swell.

"My biggest fish was a cat—weighed three pounds."

"Mine was a rainbow. Two pounds and a half."

"Say, you fellows," our friend laughs. "Come with me over here to a soda fountain. I'll treat you to some ice cream."

"Come out to our ranch if you like to fish."

"Yes, we have sunfish, perch, catfish, crappies, and if you'll go to the mountains with us we can show you some fine rainbows and brook trout."

"Maybe I'll take you up on that invitation next time I have a day off."

"We love to fish. We like to eat them, too."

"Ha, ha!" the man laughs. "So do I!"

We like that man.

ORDER OF FISHES WITH BACK FINS AND FAT FIN

(ISOSPONDYLI)

The Greek name for this order does not help us much, as it means "equal-vertebrae," or "back-bone-section-all-of-the-same-size." Regardless of the name, all of us know fish of this order because they are like trout and salmon. Salmonoid fish, they are often called. This order includes trout, charrs, salmon, whitefish, and grayling.

Between the back fin and the tail is a small fin which is soft. No other fish has this kind of a fin except catfish, and it is too well known to be confused with the trout and salmon family.

Fishes of this order are among the most important of the whole class of fish, because of their food and sports value.

Mackinaw Trout

Now let the fisherman pull and tug all he wants. His line is wrapped around a big log. Let him reel, let him try to dislodge the big trout and get him to fight again.

In desperation the fisherman jerks his rod too hard and the line breaks. He has lost his rainbow.

The swift water unwinds the wrapped line from the log. The rainbow is still hooked, though. He has ten feet of leader and line hanging from his mouth. It will be several days before fungus grows in the wound, makes the hole grow larger, and allows the rainbow to rid himself of the hated thing and be a free trout again.

Montana black-spotted and **native cutthroat** trout are somewhat alike. The black-spotted have small black spots all over the body, including the tail and back fin. The native cutthroat is thickly spotted on the tail and part of the body. The foreparts have fewer spots. The distinctive mark of all cutthroat trout is the red splash under the throat or under the gills which looks very much as though the throat had been cut. This is quite pronounced on many specimens, showing even on the gill-covers.

These are hard-fighting fish of the cold waters. They do not leap out of the water like the rainbow, but put up a good fight under the surface.

Do you think a fish could go across the Continental Divide all by itself? Look at a map of Yellowstone Park and the country south of there. Note a place called Two-Ocean Plateau between the headwaters of the Snake River and the Yellowstone River. There is a small lake there where cutthroat trout have been seen to enter from the Snake River and go out down the Yellowstone River, thus crossing the Continental Divide.

Brown trout, sometimes called **Loch Leven,** is the only true trout that spawns in the fall like the charr. Head, body, and back fin are red and black spotted. Black spots are round or X-shaped. Some spots have a pale border. Fins are usually yellowish on margins. The body color is brownish, sometimes almost black.

Golden trout are natives of California and have been transplanted into a few lakes in central Idaho. They are beautifully colored fish, thriving best in lakes high up in the mountains. Their name comes from their golden color, although this is

Eastern Brook Trout

not always very pronounced. The golden trout's general appearance is very much like a rainbow except that the parr markings or black sooty blotches on their sides are much more prominent than the rainbow and the general color is golden.

Charrs. The name is from the Irish language and means red, or blood-colored.

These are trout with scales so small that most people would say they are altogether without scales. There are other definite characteristics, too, such as the arrangement of teeth in the mouth. One great difference between charr and the true trout is that all charr spawn in the fall instead of spring.

The **mackinaw trout,** a lake fish, sometimes called lake trout, grows to enormous size. Heart Lake in Yellowstone Park has produced some as heavy as thirty pounds. Bear Lake, as well as Upper Payette Lake and Pend Oreille Lake, has been planted with mackinaw trout. Jackson Lake in Wyoming has produced forty-pound mackinaws.

Mackinaws are not active fighters, but on account of their great size, it takes a skillful fisherman to land a big one.

This fish has pale grey or whitish spots instead of red or black. Its flesh is delicious.

Eastern brook. This is the highly esteemed brook trout of the eastern U. S. It has been planted in many streams of Idaho. It is a most beautiful charr with red spots on its body. The arm, leg and rudder fins are edged in white. On the back are wavy markings in black.

More has been written of this brook trout than all other trout put together. Whenever you read a story that mentions a brook trout and its superior quality, you may be sure it is this one.

In many places in the West, the brook trout is not as well liked by fishermen as other trout. The chief objection is that it does not fight hard enough.

Brook trout quickly take advantage of beaver ponds and other small bodies of water where food is abundant. They thrive well, too, in high mountain lakes if there are a few small gravelly bottomed creeks flowing in so that they have good spawning beds.

Brook trout in Idaho generally do not grow to as large a size as other charr such as Dolly Varden and Mackinaw. In mountain streams a one-pounder is good-sized. Little ones, seven, eight and nine inches long, fried in bacon grease make a fine meal when one is on a camping trip.

Dolly Varden. Next to the Eastern brook the Dolly is the most prettily colored charr. It is found almost anywhere in Idaho waters, lakes as well as streams. The Dolly, a very greedy feeder, likes minnows, other fish, all sorts of insects and even chunks of bacon or beef. Best of all, he likes the eggs of other fish. Salmon eggs or sunfish eggs are bait for Dolly Varden.

This fish grows to good size. Eight or ten pounds is not at all unusual. However, it is not as good a fighter as some other fish, and on account of its tendency to eat and destroy other valuable fish it is not held in high regard by fishermen.

DOLLY VARDEN TROUT

Cutthroat Trout

BLACKSPOT AND DOLL

Blackspot was a beautiful trout that lived part of the time in a big river in Idaho. He weighed three pounds and was nineteen inches long. All over his upper body were small black spots. Even his back fins and tail were spotted, too. Blackspot's body was silver and his gill covers were splashed with red. He was a cutthroat trout. He got his name from two scarlet markings on his under jaw. Blackspot and his kind did not deserve to be called cutthroats any more than the salmon deserved to be called murderer, or a rainbow trout a child-eater. For Blackspot was not only a very fine-looking fellow, he was a good sport, too. He loved the cold, clear mountain water that rushed and tumbled over the rocks in the small creeks. He loved the deep pools and riffles of the river in the wintertime.

Now Blackspot lay at the bottom of a deep pool, worrying. His mate, Rubythroat, was upstream in the swift water, feeding. Ruby always became a bit restless at the time of the year, just before they began a long swim from their pool in the river upstream to where a small creek came in, and up this to small water, far back in the mountains. Blackspot had other worries besides Ruby. As he looked downstream for the hundredth time, his biggest worry moved like a shadow into the pool.

Doll was a big, fancy colored trout, two feet long, weighing five pounds. There were bright yellow and orange spots on Doll's green body. On his back were white spots instead of black ones. All this made Doll a very pretty fellow. He was a Dolly Varden trout and got his pretty name from his good looks, certainly not from his actions.

The real Dolly Varden was a beautiful girl in one of Dicken's books, who always wore very pretty, brightly colored dresses and ribbons. The real Miss Dolly Varden, "pink and pattern of good looks," daughter of the locksmith, Gabriel Varden, was a very fine girl.

Dolly Varden trout do not deserve such a fine name. Doll

is not even well thought of by other fish. He is too rough, too much of a bully. Blackspot did not like Doll at all. He did not even like the pretty colors that Doll wore. He did not like Doll's selfish, greedy ways. In fact, there was nothing about Doll that Blackspot did like, so he stayed away from the bigger fish when he could.

Every spring as just about this time, Doll showed up to bully Blackspot and his mate, Rubythroat. Doll stayed in the lower end of the pool behind the other fish. He waited to see what the two spotted trout were going to do. When Ruby come back to lie near her mate and rest, Doll made a swift rush at them with mouth wide open. He meant to drive them away.

The spotted trout fled upstream to swift water. Ruby kept on going upriver and Blackspot followed her. Every time they came to a pool and stopped to rest, the big bully caught up with them and drove them out. Once or twice Blackspot dashed at Doll in a terrible fury, trying to drive the big trout away. It never worked.

For several days the cutthroat trout played a game of hide-and-seek in the river with the Dolly Varden trout. Once or twice they lost him, but Doll always showed up to hang on like grim death. All the time the fish were travelling upstream. They passed the mouths of several small creeks. At one of these they tried a trick on Doll. The spotted trout hurried ahead and turned into one of these small streams. Doll followed them closely. He knew that they would, sooner or later, go to the very head of one of the small streams. The pair doubled back past Doll where a small island divided the stream. Back in the river Blackspot and Ruby continued on upstream. They felt good. Doll had been outwitted and was now, no doubt, searching for the pair frantically 'way upstream at the headwaters of the creek.

In a day or two, the blackspotted trout found a creek emptying into the river that was exactly to their liking. Up they went, Ruby always in the lead, a few feet, Blackspot following in order to guard against a sneak attack by Doll. Two more peaceful days passed. The pair was now away up in the forested mountains where the water was very swift. Pools were few up here and the water in them shallow.

One day Ruby began to make a nest for her eggs at the lower end of a shallow pool. She lay on her side and flapped her tail gently against the gravelled, sandy bottom. The small pebbles danced under her tail and the water washed them downstream. The larger stones, of course, were not moved at all, but the sand and gravel between them were churned out. This made some rather deep chinks down in the creek bottom. It took Ruby two hours to dig out a place large enough and deep enough to suit her.

BLACKSPOT AND RUBY IN NEST

Blackspot did not help with the digging. He stood guard below the nest. Sometimes he would move to the upper end of the pool and stand guard for awhile. It was Doll that he feared most of all, though, and most of the time he watched for him.

Ruby finished her first nest. A circular place three or four inches deep and more than a foot across was dug out. Then she started to lay her eggs in the nest. Blackspot came to lie beside her and as she laid her eggs, he spread a milky fluid, called "milt," over them. There was a slight eddy in the nest that helped the eggs to settle down in the cracks between the larger rocks, but a few washed away downstream.

After awhile, the fish left the nest. Ruby covered it with

gravel. She got upstream a few inches, lay on her side and flapped the bottom with her tail just as she had to dig the nest. The churned-up gravel and sand were carried into the nest to fill it up. She kept right on churning after the nest was filled, and until another nest was dug. By now, there was a heap of clean, washed gravel over the first nest.

Blackspot and Ruby settled into the new nest and were laying eggs and milt into it, when a dark shadow was cast across their heads. Blackspot dashed out at the intruder. It was, of course, old Doll who had at last found their nesting grounds. He retreated downstream at Blackspot's furious attack, but Doll had accomplished what he intended. When Blackspot whirled out of the nest, he disturbed the settling eggs and dozens of them floated downstream. Old Doll held himself in the riffles and gobbled up the eggs as they floated down. Then he went upstream to worry Blackspot some more and get more of the luscious golden eggs to feast upon. It was too late for the blackspotted trout to find other nesting grounds, so they stayed and fought Doll away the best they could.

No wonder Blackspot hated Dolly Varden so much.

Doll and his mate made their nests and laid their eggs in the fall instead of in the springtime, like blackspotted trout. Blackspot waited in the small stream until fall came and the Dolly Varden trout made their nest a mile or two down from the blackspotted trout's nest. He tried to use the same tactics as Doll. It didn't work. Doll chased him so furiously that only a few Dolly Varden eggs found their way into Blackspot's stomach.

Blackspot drifted downstream to the big river early. He was not very happy about the way Doll had feasted on trout eggs and himself not getting many of the Dolly Varden eggs. So Blackspot brooded in the deep pool at the bend of the river.

Every day Blackspot would see a fisherman come to stand near the pool and cast his lures on the water. Oftentimes, the shadow of the fisherman in late evening would fall across the water before the man himself came to stand on the high bank and cast his fly. Blackspot was never fooled. The pretty colored flies that alighted so softly on the water, yellow and black and orange, reminded Blackspot of his enemy, Doll, and he just sulked all the more. The fisherman could see the big cutthroat, and he tried all the harder to get it to rise and strike his hook. The fisherman tried all the flies in his book, but to no avail. Then he tried bait, grasshoppers and even angleworms. By this time, Blackspot would not have taken a bite at anything this shadow-making creature offered. There were deep, cruel scars in the trout's mouth where barbed hooks had pulled through his flesh. He wanted none of the things a fisherman offered him.

Then Doll came, Doll the big fellow that Blackspot hated! Doll was fat and as much of a bully as ever. He stayed in the pool with Blackspot. When the spotted trout went upstream to the riffles to feed, Doll went, too. Many times, as Blackspot was just about ready to gulp down a crayfish, small whitefish or black fly, Doll would rush in, shoulder Blackspot to one side and take the food. A half mile upstream was a small bed of freshwater shrimps on which the cutthroat liked to feed. Doll always followed him and shouldered him clear out of this prize feeding grounds.

No wonder Blackspot sulked in the deep pool between feeding times.

Doll would not rise to the fisherman's lure, either. Even when the hook was heavy with a fresh, juicy grasshopper and weighted so that it sank right in front of Doll's nose, he would not bite. Doll too had some scars on his mouth where steel hooks had torn through his skin and flesh. He was a hard fish to fool. Still, the fisherman came every day or two.

One bright, warm fall day, there was a sort of vague unrest in Blackspot's mind. It was not like the urge to move upstream with Rubythroat in the springtime. It was not a new feeling of malice toward Doll. It was not hunger or a desire to move to a new pool. Somehow or other though he wanted action. He wanted to dash swiftly, suddenly, furiously at anything that moved near him. When a tiny midge alighted on the water surface, Blackspot darted upward like a flash and struck the little fly. A floating chip got the same treatment. Now a chip of wood was not good food, yet Blackspot struck it furiously, and gulped it into his stomach. The fisherman came along and dropped a small, grey-colored lure on the water. Blackspot rose like a demon and struck the lure. Before the astounded fisherman could realize that at long last the cutthroat had struck for him, the fish felt the tiny barbed hook in his mouth and quickly spat it out.

Doll watched the other trout's antics. Doll thought that Blackspot must be getting some tasty bits of food. The next cast the fisherman made, Doll was waiting and he struck, too. The two fish clashed at the surface of the water and made a great splash.

The fisherman was now hopping up and down with excitement. Not only was the big cutthroat striking, but there was a much bigger Dolly Varden striking with him. Again the fisherman cast and again two big fish struck. Neither were hooked.

Doll was excited, too. Rarely had he ever been bested by the cutthroat. Next time! Next time the hook came down it was baited with salmon eggs. Ah! Salmon eggs! Just as good as cutthroat eggs to Doll. He went furiously after them.

Was Blackspot tired, or did he purposely rise rather weakly toward the hook laden with salmon eggs? Anyhow, Doll got there first by a good two feet and greedily gulped down the bait. Blackspot went back to lie on the bottom and watch the struggle between Doll and the fisherman. It ended when the fisherman finally pulled Doll out on a sloping sand bar 'way down the river.

Blackspot took a great gulp of oxygen-rich water and let it ooze slowly through his gills. Now he was content.

Little Redfish, Sockeye or Blueback Salmon

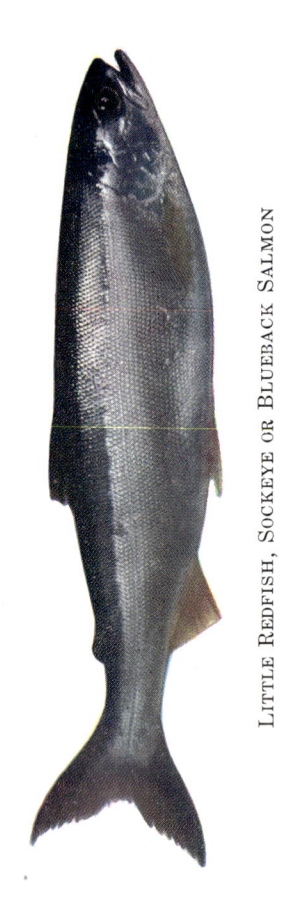

THE SALMON FAMILY

Now we come to the **salmon,** the Pacific Ocean fish who live most of their lives in salt water, but come back to fresh water to lay their spawn and die.

There are several kinds: **chinook** or **king salmon, sockeye** and **steelhead.** Sockeyes have a deeply notched tail fin. The name sockeye comes from the Indian *saw-qui* and has nothing to do with the fish's eye.

Salmon eggs are laid in gravel beds far up on the headwaters of the Columbia River. In Idaho the main salmon streams are the Kootenai, Clark Fork, Clearwater, Salmon, and Snake Rivers. Salmon River is counted as the greatest spawning grounds of sea-run fish in the whole United States.

Tiny salmon are called "parr." Black stripes across their bodies up and down are called parr marks. As soon as they outgrow their parr marks they are known as "smolts." Smolts go down to sea at the age of one or two years. In the sea they are called "grilse" until fully grown, at five or six.

Pacific Coast salmon die soon after the eggs are laid in the fall. They never have a chance to see any of their children. The children are born after the old ones die, so none of them ever sees one of his parents.

Salmon is one of the valuable food supplies of the West. Lakes Pend Oreille, Coeur d'Alene, Priest, and other big lakes furnish fine salmon fishing in addition to that in the streams.

Because of the habit of salmon returning to the streams where they were born to lay their spawn, it is necessary to plant young salmon from a hatchery into whatever streams it is desired to have them return to. If all hatchery salmon were planted in one stream, all of them would return there. None would seek other streams.

Some salmon have been kept in lakes and streams, where there was no outlet to the ocean, for so long that they do not have the habit of going out to sea. These are called "landlocked" salmon. The saw-qui is one of these. After losing the sea run habit, saw-qui can be raised in streams that do go to the ocean and still stay land-locked. Pend Oreille Lake, Alturas,

Soon the man is reeling it in, hand over hand. Goodness, is he ever going to get to the end of the line? Are we never going to see the fish? What if it gets away after all?

Another man moves over with a gaff. This is a sharp iron hook on a wooden handle. He stoops over the side of the boat. After what seems to be an hour of waiting, the chinook salmon at last breaks water in a great leap. We see that it is a big one. Down he goes again to struggle under water against the steel hook that is dragging him where he doesn't want to go. Again he leaps and dives. The man pulling on the line takes plenty of time. He doesn't know how excited we are. At last he pulls the salmon's head out of the water against the boat and the other man strikes the iron gaff into its head and pulls it aboard to flop around a few minutes before it dies. We admire the beautiful fish. It is a dandy—weighs all of thirty pounds, the captain says.

"Number four!" shouts the captain and follows with, "Number one! Number seven!" Ah, now we're getting into good fishing grounds. "Number eight!"

Soon all of us are busy pulling in fish and dropping the lines out again. It is mighty exciting for awhile. Soon we begin to get tired. What do we want with all these fish, anyway?

"What is the limit, Captain?" we ask, hoping that we haven't broken any game laws.

"There is no limit. Take all you want."

"We've plenty now. Take us back to shore where we can give away some of these."

"O.K.," says the captain of the *Royal King*. "We're out about fifteen miles, but we can make it back in about an hour."

We can see no land at all at first, but soon the dark purple hills show up and grow larger and larger until we run into the little harbor again. We give four fine salmon to some tourists who do not have time to stop and fish. Then we go on our journey and finally arrive at Challis, away up in central Idaho. There are salmon here in the Salmon River. We have seen them many times, and they look something like the salmon we caught in the Pacific Ocean. Surely, though, Pacific fish would not come all the way to upper Salmon River! That is

Chinook Salmon

too long a trip, we should think. The salmon don't think so, for all little salmon are born in fresh water as far up on the heads of streams as they can swim.

This is the way salmon live!

A THOUSAND MILES DOWN TO THE SEA

Chinook was born in Bear Valley Creek at the headwaters of Salmon River, far back in the mountains of Idaho. As a tiny fellow, hardly an inch long, he swam with hundreds of others among the rocks and pebbles with a yolk sac hanging to his belly. It was from the egg yolk that Chinook got his food to grow.

After a few weeks the yolk sac was all gone, and Chinook had to find other food. He swam and swam in the shallow water looking for food. He found the eggs laid by a mosquito and greedily downed them. Later on, the eggs of some other insects floated down and the baby salmon captured them, too.

Little Chinook lived in the headwaters of Salmon River for a whole year. Then one day he felt an urge to be gone. He felt like travelling, just as everyone does on a fine spring morning. At first he just wanted to go farther downstream where the pools were deeper and the riffles stronger. He wanted to try out his strength in bigger waters. There was more food in the river, too. There were small whitefish to catch, big May flies and often a crayfish. No longer did he depend on the eggs of mosquitoes and flies. Those were baby food, and Chinook was getting to be a big fish.

Chinook liked deep water. It gave him room to move about. He was now nearly a foot long, a big, powerful, independent fish of the river, looking for even bigger waters.

In the Columbia River, Chinook thought at first that this was where he would live. It was wide, long and deep. He stayed in it one whole summer, then moved on. There was something he could not resist, drawing him on to the biggest of all waters, the wide, boundless ocean. There were pleasant booming sounds against the sides of his body, sounds of tireless waves breaking against the rocks. He went out to meet them. He did not exactly like to live in salt water at first and stayed in fresh water where Columbia River emptied into the Pacific until he got used to it.

Once accustomed to salt water, he loved it. There was an abundance of food, and water so deep that even Chinook dared not try to find the bottom. Life here was almost a dream. True, there were monstrous fish in the ocean that could gulp Chinook down in one bite if they caught him. There were great sea lions that lived on the rocks along the coast who would gladly eat Chinook, but they could not catch the swift salmon that had come down from the high mountains, a thousand miles away. There were devilfish, seals, sharks and whales. There were fishermen, too, as Chinook learned to his sorrow one day, when a bright, golden, swiftly moving spinner caught his eye. He struck at it hard and found himself caught in the jaw with a steel hook on the end of a long, stout line.

Chinook fought with all his might. He leaped clear out of the water, once, twice, three times. Still the hook held. He tried to spit it out, as he did the claws of a lobster or crab, but with no success. He dived down deep, shot like a bullet toward the surface, then whirled at top speed—there was a tearing of flesh and sinew, a terrible pain in his jaw, but Chinook was free. The hook had torn out. Deep on the ocean floor, Chinook lay while his jaw healed. Ever afterward, he ignored the brightly flashing spoons of the fishermen. He knew they meant torture and pain.

Chinook was six years old when he felt the urge to move back to fresh water. This time he heard other sounds against his body, sounds of water tumbling over rocks and of swift currents dashing against wooded shores. He smelled fresh water, too, water without the tang of salt. For four years he had lived in the ocean. He had eaten greedily of the food that was so plentiful. Now Chinook was three feet long and weighed thirty pounds. He had roamed hundreds of miles in the ocean, south, west and north without ever once coming to fresh water. His longing now was for the Columbia River and he gradually moved that way. Chinook knew where the mouth of the river was from its sounds and smells. He knew where he wanted to go, just as the honkers know where to go when the right time comes. There were many other freshwater rivers emptying into the Pacific Ocean, but only the Columbia would take Chinook where he wanted to go.

Fresh water did not ease the itch to travel, but it slowed him down. For several days he lurked in the shallow waters at the mouth of the river with hundreds and thousands of other salmon, all with the same thought in mind, to go upstream to small water.

Chinook did not know it, but there were a great many more difficulties to going upstream than there were coming down four years before. It is always easier to drift downstream than it is to swim back up. But even if Chinook had known of all the troubles ahead, he would not have turned back to the ocean. Returning to where he was born was a part of his life, and he could not avoid it any more than he could have avoided being born in the first place. He had to go. He was a salmon.

The first danger Chinook encountered was nets. For several miles up the river nets were stretched out to catch the salmon. There were hundreds and hundreds of them. It did not seem than any fish could swim past them. Many fish were caught for the canning factories ashore. The nets were often so heavily laden with salmon that horses were used to pull them in. The horses were kept in funny-looking stables, built high up on timbers, above high tide. There were sloping runways down from the stables to the sand bars that were exposed at low tide. Boats brought the nets to shallow water where the horses were hooked on. Sometimes the nets were so full that even horses had a stiff pull getting them ashore.

Chinook came to a net the first day he was in the river. There were a few fish caught in it, and he circled around the end to avoid it. Then he swam very deep down, almost on the bottom of the river and went under many of the nets. At a sand bar he came up and swam smack into a net! A cord became fastened firmly in one of Chinook's gills. He twisted, turned, churned ahead, but struggle as he might, he could not get loose. Dozens of other fine salmon were caught, too, and soon the boatmen came to pull them to shallow water where the horses could be hitched, one at each end of the long net, to pull it ashore. The fish fought desperately. On the beach out of the water, they flopped in the sand, gasping wildly for breath. Fishermen came along with iron pipe to knock the captured fish senseless.

The river was rough that day. A wind blew in from the ocean and little waves pounded at the sandy beaches. One white-capped wave, a little bigger than the others, came rolling into the sand bar where Chinook had been fighting his best to get back into the river. He lay on his side, gasping. The little wave touched him. He gave one mighty flop and slid back into the water. A fisherman came running—but he was too late. Chinook darted away at top speed.

After that narrow escape, he travelled very carefully, sort of feeling his way along, until he was past all the nets. Then he ran into a new danger. Hundreds and hundreds of fishermen sat in boats trolling with brightly polished spinners. Many times Chinook saw these shining things whirling through the water. He knew them for what they were, a snare for fish, and did not strike at them. In fact, Chinook was not hungry. He had not been hungry since leaving salt water. There was plenty of food in the river, but he simply did not want to eat.

There were many smells in the great river; the acid smell of rotting wood, the strong smell of thousands of logs that lay rafted along the shores, the turpentine smell of sawdust and of chemicals dumped into the water. There was the mud smell of dredges stirring up the bottom, of fresh soil washed down from plowed fields. Ashes from forest fires sifted into the stream, and along one shore was the strong stench of a city's sewage, mingled with the other bad odors. Chinook was glad to get past the mouth of the Willamette River and into cleaner water.

Many strange sounds, too, struck Chinook along the sensitive body line. Sounds came to him of chugging motor boats, of great ocean-going liners and of rafts of logs, bumping together. This part of the Columbia River was not a pleasant one for him.

A few days' travel from the nets and fishermen, Chinook came to the great Bonneville Dam. Here were thousands and thousands of salmon that had some way or another escaped the fishermen only to run up against an impassable wall across the river. Some water came over the top of the dam, but try their best, none of the fish could jump or swim over. Then they found a way. Chinook made just one little jump that took him into a small concrete pool. Water flowed into it from above. Another easy leap took him into another pool, then into a third, on and on, just like a stairway, except these were pools, one above the other, with water flowing down, instead of steps. Chinook jumped up from one pool to the next rapidly, almost as easily as he would have gone up a steep cascade. It was a fish ladder, made by the men who had built the dam so that salmon could get over the barrier. Chinook made the ascent with hundreds of other salmon. The waters above the dam were quiet, and all the fish got a few days of rest and quiet.

Rough, fighting water at The Dalles came next. Here were more fishermen, dozens and dozens of them, fishing from the rocks with dip nets. Some of them stood on platforms let down from the top of the rocks with ropes. It did not seem possible that a fish could fight his way through these roaring, foaming, tumbling waters and escape the fishermen's nets. Chinook twisted, darted, leaped and dashed his way through. Once a net almost got him. As it raised out of the water, the salmon balanced on the rim for an instant. If he had fallen forward, he would have been caught in the net. Luckily he flopped backward into the water and escaped upstream.

There was easy going above The Dalles. At the mouth of Snake River, Chinook turned. The Columbia was still the larger river. Why did Chinook choose the Snake? Was it because four years before he had come down the Snake River, headed for the ocean? Yes. It was because the Snake was the way for Chinook to go. Like other wild things, he knew the way. Again, at the mouth of the Clearwater River he chose to stay in Snake River. When he came to the mouth of Salmon River, he had to choose again. This time he quit the Snake and took the Salmon, a clean, cool, sweet-smelling river, with only the sounds of swift water rushing down between mountain cliffs.

There were other salmon going upstream, and as the river grew smaller some of these fish fought viciously with Chinook. Every day he had several fights and in a few days was badly scarred and hurt. He kept going on up, drawn as if by a magnet, to Bear Valley Creek at the very headwaters of Salmon River. That day he found his mate, a beautiful sleek salmon like himself, who had managed to elude the fishermen and who had come back to where she was born. Here where the water was hardly deep enough to cover their backs, she dug her nest and laid her eggs.

Chinook fought away other fish, his whole time taken up

in guarding his mate while she prepared the nest deep in the gravel at the bottom of the stream. She laid her eggs and Chinook spread his milt over them. She covered them with gravel. They would hatch next spring into tiny salmon.

Now a great wave of fatigue swept over Chinook and his mate. They were weary. They had no ambition to travel, to seek food, to do anything. Chinook's body was bruised, torn and cut from fighting. Slowly they drifted downstream from the nest. They had not eaten for weeks.

Their lives had been lived. They were through. They had been a thousand miles downstream and a thousand miles back. There was nothing more for them to do. As naturally as they were born, the old salmon died.

At the head of the stream, covered with sand and gravel, were ten thousand eggs, laid by Chinook's mate. Barring accidents, they would hatch out into ten thousand tiny salmon, ready and eager for their own thousand miles down to sea—and back again.

WHITEFISH

Whitefish are small-mouth, light-colored fish of the cold mountain waters. They reach their largest size in the lakes of Idaho, growing to three or four pounds in weight.

WHITEFISH

You will never mistake a whitefish if you look for all the fins that a trout has and note the very small mouth. It is a good fish to eat, particularly if you catch it in winter. It lies on the bottom of deep pools so it is necessary to use bait and a sinker to catch it. In summer it will strike at a fly, like trout. It is not as great a fighter as a trout, but mighty good to eat just the same.

Another species found in Lake Pend Oreille is called the **Lake Superior whitefish.**

GRAYLING

In the **grayling** we have a most beautiful fish, sometimes called the "flower of fishes." Some of the natives of the Far North say that the reason the grayling is so beautiful is that he feeds on gold.

The grayling has an enormous back fin which has been likened to a peacock's tail, because of its size and brilliant colors. In general, a grayling's body is light colored, a sort of purplish grey. There are blues and bronzes, blacks and silvers, roses and pinks, and even greens in the grayling's color scheme.

In action, this fish is graceful and swift. Many fishermen say it is equal, if not superior, to the rainbow in fighting spirit. Often the grayling breaks water in a long leap through the air in his efforts to break loose from hook and line.

Grayling

The grayling's mouth is small, but not as small as that of the whitefish.

This is a fish of Arctic regions. It thrives only in the coldest waters. Those found in Idaho have been transplanted from Montana and are known as Montana or "mountain" graylings. The headwaters of the Missouri River are the native home of this fish.

If you should go into a fish hatchery and see some tall jars full of dancing, very small eggs, they will be grayling. The eggs must be on the move all the time or they will stick together and die. Water bubbling up from below in the glass jars and flowing out of the top keep the eggs in constant motion. Young graylings are difficult to rear in the hatchery, one reason being that their mouths are too small for ordinary hatchery food.

We wish there were more of these fine fish.

ORDER OF SUCKERS AND MINNOWS

(PLECTOSPONDYLI)

FAMILY OF SUCKERS   Catostomidae
FAMILY OF MINNOWS   Cyprinidae

Suckers are bottom feeders. They eat vegetation and sluggish animal life.

Minnows are like mice, found anywhere. Their greatest value is as food for more desirable fish.

SUCKERS

Suckers have a small mouth on the under side of their head which gives them the Greek name of *Catostomidae*, meaning "downward-mouth-family."

SUCKER

Suckers have arm, leg, rudder, back and tail fins, but no fat fins. They are not considered a very good food fish, although in some waters they are not bad. Indians used to like suckers because they were so easy to catch. One reason most people do not enjoy eating suckers is because they have so many small bones in them.

Suckers are easily caught with a weighted line and small hook baited with an inch of angleworm. Often they become so numerous that seining is necessary to get them out of a lake or stream in order that other fish may have a chance to live.

MINNOWS

The first thing to know about the minnow family is that they are not necessarily small fish. Many of them never grow to a size more than six or eight inches long, while others, like the German carp, may grow to be two or three feet long and weigh fifty pounds.

The brightly colored gold fish in your bowl at home is probably a member of the minnow, or carp family. The Greek name is *Cyprinidae*, which means "carp-like" family.

Often minnows are called "forage fish," because they are food for larger fish. The back fin of minnows, which in some other fish is very stiff and dangerous to anything swallowing it, is short and soft. Often fish can swallow minnows without the least fear that sharp spikes will lodge in their throats. The fat fin is, of course, absent.

SQUAWFISH

The **squawfish** is one of the best-known members of the minnow family. They often grow to a length of two or three feet. One foot is a common size.

They are large-mouthed, rather ugly fish of poor eating qualities. The name "squawfish" comes from the ease with which Indians caught them in traps, woven from willows.

Squawfish eat food needed by trout. They are very destructive to the young of game fish and are not gamy fighters.

Some of Idaho's lakes and streams have altogether too many of these fellows.

Carp. This biggest of all minnows was introduced into western waters from Germany, where it is highly regarded for its food qualities. Here it has become a nuisance, an undesirable, out-and-out pest. It is a bottom feeder, rooting in the mud for tender vegetation and as much small animal life as it can find. It keeps the water muddied all the time, besides eating any bit of food in it. Even ducks, as good as they are in finding food, get nothing from a carp pond.

CARP

Carp are egg eaters, robbing any fish nest they find. They are too slow in moving about to catch other fish. When winter comes, they lie on the bottoms of sluggish waters, more or less buried in the mud.

One carp may lay as many as a million eggs, a hundred times as many as a big trout. Carp are too lazy to build a nest.

Dace. This is a small minnow growing only to about two and one-half inches. It is a mottled black, darker than the redsided minnow. A dark band runs along its sides.

This minnow lives in swifter water than the shiner and is usually found in swift streams where trout may be. It is used as food by the larger trout and other bigger fish.

The food of the black-nosed dace is mostly tiny animal life while they are young. Later most of their food is black fly larvae, May fly nymphs, other insects and some plant material.

SHINER

Red-sided minnow or **shiner.** This is a minnow that grows to four or five inches in length. Its name comes from the male fish which has bright scarlet sides during the spawning season, which is early summer. The female fish is a dull red. During the rest of the year they lose most of their red and are silver colored.

The shiner lives in warmer, quieter water than the trout and is usually found in lakes, ponds and small creeks, with suckers, bass, and bullheads. It spawns in the riffles of streams.

Utah chub and **Columbia River chub** are longer, commonly

CHUB

growing to a length of twelve inches. They travel in schools of fifty or more, usually in large streams. This species is too bony to be eaten; in fact, few of the order of minnows are good eating.

Stonerollers. This is a species that goes up the fastest streams in summer and returns to quiet waters for winter. It rarely grows to a length of six inches. It is food for trout. Being of a tough, hardy nature, stonerollers are much used as live bait for bass.

ORDER OF CATFISH

(NEMATOGNATHI)

Fish with whiskers, a big head, enormous mouth, and spiked arm fins belong to this order. The Greek name means "threads" (or hairs) on the jaw. All are scaleless. They have a fat fin and a very small back fin.

The only family represented in Idaho is *Siluridae,* denoting simply "river-fish-family." None of its members is native to this state or anywhere else on the Pacific Ocean drainage. Three kinds have been introduced; the **horned pout, bullhead,** and **spotted channel.**

Horned pout and bullhead grow to a size of sixteen inches and a weight of nearly three pounds in the warmer and more sluggish waters of Idaho. They have a very large mouth, are dull brownish and blackish colored, and the tail is not so deeply notched as that of other catfish.

BULLHEAD

Bullhead can live in muddier water than any other fish unless it is the carp. A bullhead can live on dry land longer than almost any other fish. They can bury themselves in mud bottoms and live for days and weeks after the water is all gone.

Channel Catfish

They build nests for their eggs under logs or stones near the shore. The eggs require constant attention. Both parents stand guard and take care of them. The old fish take the eggs in their mouths and blow them out again, stir the nest with their whiskers, and fan the eggs around with their fins to keep them from smothering. Even after hatching, the parents must mouth, stir and fan the little ones for seven days. The old ones guard the school for some time after that. One stands a close guard, the other a distant guard. If an enemy approaches, the parents rush in and muddy up the water so that the young ones cannot be easily found. Any young ones that stray from the school are punished severely by their parents. They are eaten alive!

Bullheads are not considered a very gamy fish. They are usually caught with bait, such as worms and little chunks of meat. Sourdough balls are used in some places with good success.

The flesh of catfish taken from reasonably clean waters is very good.

The **spotted channel catfish** is one of the best of the whole family and has been transplanted to a few Idaho waters where it is given protection in the hopes that it will multiply. These grow at times to fifteen pounds, although five pounds is a big channel cat.

These fellows are light olive or bluish on the back, paler on the sides and white on the belly. Black spots are scattered irregularly over the body. The mouth is smaller than the pout's. Eyes are small and low down on the side of the head. Tail is sharply forked.

Channel cats prefer cleaner, swifter waters than the pout. Their flesh is firmer and better flavored.

They should do well in the Snake River and other waters of this state.

ORDER OF SCULPINS

(SCLEROPAREI)

These are sometimes called **mountain bullheads**.

Look under the flat stones in the creek or river. If you see some tadpole-shaped fish scuttling away you may know they are sculpins, sometimes called bullheads, or Miller's thumb. Now get a kitchen fork with sharp tines, to spear them with, because it is hard to catch them in your hands. Raise a likely rock. The little two- or three-inch-long sculpins won't run for a few seconds, and this gives you a chance to spear them.

Sculpins lay their eggs on the underside of stones where they hang like a cluster of grapes. The young seek shelter in shallow water where other fish cannot get to them. Practically every fish in the streams preys upon sculpins. They are food for just about all other animals big enough to catch and eat them.

In deep ocean waters there are many families and species of the order of sculpins, some of them of great size. But these do not come to fresh waters for us to see.

ORDER OF STURGEONS

(GLANIOSTOMI)

These are fish with rows of large heavy bony plates instead of scales, and mouths with "barbels." Barbels are long, slender, fleshy hair-like growths that look like a beard.

Only one family is represented in the fishes of Idaho, *Acipenseridae* and one species, the **white sturgeon** of the Columbia River drainage. The skeleton of these sturgeons is of soft cartilage, instead of hard bone, hence the Greek name which denotes "soft-skeleton-family."

Sturgeon grow to great size. One 8 feet long and weighing 500 pounds is not at all unusual for the Snake River.

These fish feed on the bottom. They have tube-like mouths which suck in smelt, lamprey eels, other small fish and food which they can find.

The meat of sturgeon is very good but it is their eggs which give them the greatest value. Caviar, which sells for a very high price, is nothing more than salted sturgeon eggs. When one is caught before egg-laying time the eggs, or "roe" as the egg mass is usually called, is carefully saved. They are first rubbed through a screen to separate the eggs, salt is added to make a brine. After the roe is thoroughly salted it is taken out and drained through a sieve, dried and canned. Now it is a highly expensive food, one which only wealthy people can afford.

Sturgeon are becoming very scarce. They are practically extinct on the Atlantic Coast. The Snake River and the Columbia River contain nearly all the remaining sturgeon in the United States. It is said that it requires twenty-one years for a sturgeon to grow to maturity. This valuable food fish should be protected.

STURGEON

REARING YOUNG FISH

Why have fish hatcheries? Fish know how to make nests and lay their eggs in gravel at the bottom of a stream, so why not let them do it?

The best way to answer these questions is to ask some more.

Why have hen hatcheries or hen houses? Hens know how to make nests and lay their eggs out in the fields and sagebrush, so why not let them do it?

We have hen houses to protect the eggs from hawks, crows, skunks and other egg eaters, also to protect the tiny chickens from their enemies and to see that they are well fed. Oftentimes, the eggs from hens are collected and put in a huge incubator instead of allowing the hens to set on them. In this way, two or three hundred chicks are hatched at the same time. There are hardly any losses.

Mother fish do not sit on their eggs to make them hatch. Most of them just lay the eggs in the creek, cover them with gravel and go away. All egg-eating animals have free access to the fish's nest and often eat nearly all of the eggs. Then, too the baby fish have a hard time finding food for a long time after they are born. Many of them are eaten by larger fish and other animals. All in all, if ten eggs out of each hundred hatch into fish and grow to be two or three inches long, the mother fish is lucky. Only one or two out of each hundred grow as large as six or seven inches long.

Now in a hatchery, eighty eggs out of each one hundred hatch into fish. And ninety out of each hundred of these fish will grow to be two inches long. If there are good rearing ponds at the hatchery, we can expect nearly all of these to grow to six inches long.

So you see that, while we might get only four or five fish out of each hundred eggs laid by the mother fish in the creek, we get about seventy fish out of each hundred eggs when we take care of them at a hatchery.

How in the world do you get the fish's eggs to put in the

hatchery? Do you collect them out of the fish's nest in the streams?

We do better than that. We collect fish eggs direct from the fish themselves. Take an old mother trout, for instance; it isn't at all difficult to get her to give us her eggs.

All of us know what trout eggs look like. They are round and about as big as a capital O. Their color is a beautiful orange. No, they do not have a hard shell like a hen's egg. The shell is soft. Another name for fish eggs is spawn. A big mother rainbow trout may have as many as three thousand eggs. Rainbow trout get ready to lay their eggs in the springtime, along in June in the high mountains. In lower altitudes they sometimes spawn as early as February. They go up the small streams looking for good places to make their nests. They come to a trap that the hatchery men have built across the stream and are caught in a small pen. In a few days the pens are so full of fish that they cannot move about. They are packed in like a crowd at a football game.

Each egg taker wears rubber boots on his feet and a canvas glove on his left hand. He grasps the fish around the small part, just ahead of the tail. The glove is to keep the slippery fish from wriggling out of his hand. Then he places his right hand under the fish's head and lifts it from the water, holds it over a shallow pan, moves his right hand down along the sides of the fish with a very gentle pressure and the eggs squirt out of the fish's body in a golden stream. Sometimes the egg taker applies his right hand several times in the gently squeezing motion in order to get all of the eggs. The mother trout is then allowed to continue upstream, unharmed.

As soon as the bottom of the pan is covered with eggs, milt from the father fish is spread over them. This is a white, milklike substance which is taken from the male fish in much the same way that eggs are taken from the mother. The father trout is tossed over the trap and follows the mother upstream. The eggs are stirred by hand to be sure that some of the milt touches every egg. All eggs not touched by the milt later turn white and will not hatch. More eggs are stripped into the pan, with more milt, until it is full. The eggs are washed clean in two or three changes of water and placed in a large can to harden. After standing for fifteen to twenty minutes they are ready to be taken to the hatchery and placed in wire trays and baskets.

Now, let's visit a hatchery.

It is a cool, shady place with water running out of pipes and splashing down into wooden troughs to flow out at the lower end through a screen. We have already learned that a trout's eyes have no eyelids and that they must suffer in bright sunlight. Direct light will kill an egg in two minutes, so shade is necessary. Water absorbs plenty of oxygen as it splashes through the air into the troughs. It must be clean water as any muddy or sandy sediment will kill the eggs. Sudden jars, or rough handling, will kill them, too.

In one trough the newly laid eggs are placed. These look just like they did when first taken from the mother trout except that they are larger. There are a few white ones. These are dead and must be picked out, or they will cause others to die and turn white. In the trough that stands next to this one are some different-looking eggs. These are fifteen days old; and the two black spots you see in each egg are eyes. These are called eyed eggs.

Moving over a few troughs farther, we find some eggs that clearly show the outline of a small trout curled up inside. They are thirty days old.

As we go to some other troughs, we find tiny trout swimming around with a round sac attached to their bellies. This is part of the egg and is called the yolk sac. The tiny fish gets its nourishment from this sac. It is now fifty days old and will carry the yolk sac for twelve days. After that, it must find its food and eat like any other trout.

Fifty days for trout eggs to hatch into trout? Why, hen eggs hatch in twenty-one days.

That is right. If you could watch what goes on inside a hen's egg you would see a chicken mature something like the fish matures; and in twenty-one days. The hen keeps her eggs at an even temperature of about one hundred and three degrees warm by setting on them. The trout eggs were in water of only fifty degrees temperature. In water fifty-one degrees, the eggs hatch in forty-five days; where the water is fifty-two, it takes only forty days. Colder water takes longer. Forty-nine degree water takes fifty-five days and forty-eight takes sixty days.

Each degree difference in temperature makes a difference of five days in hatching.

Fifty-six degrees is about as warm as hatching water should be, forty degrees about the coldest that could be used to get fair results.

Step over the end troughs—they are full of tiny trout about an inch long. These are called "fry" and are three months old. The hatchery man feeds them very finely ground food and they grow fast. A chunk of liver the size of a large marble is a day's ration for a thousand fry. In a few months, they are three inches long and quite active. Now they are "fingerlings" and can be put outside in the large ponds. Here they are fed food a bit coarser than they had as fry, and, of course, there is a lot more of it.

There is a cold-storage room in the hatchery where liver and other fish foods are kept. Then there is another room where electric meat grinders cut up the food so that the little fish can eat it.

Sometimes fry are taken direct from the hatchery and planted in the streams. These tiny fellows must find food and keep away from their enemies if they are to live.

Other times fry are kept at the hatchery until they become fingerlings before being put into streams or lakes to look out for themselves. These larger fish can, of course, swim better than fry.

If the hatchery has plenty of rearing ponds, the young fish may be kept until they are six inches long, or even longer.

The hatchery does a good job of raising fish. We can easily see that. They get seventy to eighty fingerlings out of every hundred eggs. We are satisfied with the job the hatchery does, but how about getting these delicate baby fish moved to their homes in the creeks, rivers and lakes? Can that be done easily?

This is not too easily done. There is a chance for a big loss in planting fish.

Suppose we have eight thousand fry to plant in a mountain creek a hundred miles from the hatchery, and it is a hot day in August. The hatchery has a big tank truck especially made in which to haul fish and they have men who know all about how to take care of the fry. This big truck cannot go where we want to take our eight thousand fry, which the Fish and Game Department said we could have, so we will haul them ourselves in a light pick-up truck. The fisheries men tell us how to do the job and we go ahead.

We get four big milk cans—clean, of course—two water pails and a long-handled dipper. We get a box full of ice, too, because our little fry cannot stand too much heat. The hatchery man fills our cans with water and measures out the fry.

Measures? There are to be eight thousand fry. Must he not count them?

That would be a long, tedious task. The hatchery man did count fry once. He counted and found that a pint measure held a thousand fry, a little over an inch long. After that he just dipped them out of the troughs by the pint instead of counting. Now he puts two pints of the little fellows in each can, eight pints in all, which makes eight thousand fry. Quite easy, isn't it?

Then he puts a small chunk of ice in each can and we start out.

What must we remember?

First that trout must have plenty of oxygen in the water.

All right, we use the long-handled dipper as our truck rolls along the road. We dip out the water and pour it back in the can. The water takes up oxygen from the air. So we dip and pour—dip and pour—dip and pour—all the way.

Another thing to watch is the temperature. Over fifty-six is dangerous. The little ones will begin to turn their white bellies up and die. All right, keep some ice in the cans. Put in a few very small chunks at a time. That will keep the water cool.

We lose some water from the cans. Some splashes out when the truck bounces over rough places in the road. We spill some while dipping and pouring. *Stop the truck!* We must fill up!

The truck stops on a bridge. We take our two pails and fill them from the stream. If the water is warm, we put ice in it before pouring it into the cans. We must be very, very careful not to change the water our fry are in from warm to cold, or from cool to warm, except just a tiny bit at a time. We use a thermometer constantly to be sure there is no sudden change

in temperature. If we pour a lot of warm water into the cold water of the cans, there will be many white bellies floating on top. So we put in cold water a little at a time.

Finally we get to the creek where our plant is to be made. Shall we just empty the cans into the creek, toss them into the truck and go home?

No, sir! Find a shallow, shady, quiet place along the shore. We take the temperature of the stream and slowly equalize the water temperature in the can with the creek temperature. This allows a gradual mixing of the water in the can with that in the creek. We prop the can so that it will stay in position and find another good place for another can. We won't hurry. It has taken months to raise these babies in the hatchery. We can take a few hours to introduce them to their new homes. In an hour or two we come back to the first can and find that all the little fish have swum out and are already scattered in the pool, looking for food.

There are only a dozen or so that turned up their white bellies and died. We have done a good job. Now it is up to the fry themselves to evade their enemies and find food to live on.

We hope they will grow into fine, big fish.